RESEARCHES IN MAGNETO-OPTICS

MAXWELL PRESS

RESEARCHES IN MAGNETO-OPTICS

P. ZEEMAN, SC.D., PH.D., D.SC

Winner of Nobel Prize for Physics

Professor of Experimental Physics,
The University of Amsterdam

MAXWELL PRESS

Chennai Trichy NewDelhi

MAXWELL PRESS

This edition has been published in india by arrangement with Carson Books, UK

ISBN 978-93-90063-74-1 Maxwell Press

No. 44, Nallathambi Street,
Triplicane, Chennai 600 005

MJP 918

Publisher : C. Janarthanan

Publisher's Note

The legacy of a country is in its varied cultural heritage, historical literature, developments in the field of economy and science. The top nations in the world are competing in the field of science, economy and literature. This vast legacy has to be conserved and documented so that it can be bestowed to the future generation. The knowledge of this legacy is slowly getting perished in the present generation due to lack of documentation.

Keeping this in mind, the concern with retrospective acquiring of rare books has been accented recently by the burgeoning reprint industry. Maxwell Press is gratified to retrieve the rare collections with a view to bring back those books that were landmarks in their time.

In this effort, a series of rare books would be republished under the banner, "Maxwell Press". The books in the reprint series have been carefully selected for their contemporary usefulness as well as their historical importance within the intellectual. We reconstruct the book with slight enhancements made for better presentation, without affecting the contents of the original edition.

Most of the works selected for republishing covers a huge range of subjects, from history to anthropology. We believe this reprint edition will be a service to the numerous researchers and practitioners active in this fascinating field. We allow readers to experience the wonder of peering into a scholarly work of the highest order and seminal significance.

MAXWELL PRESS

To the Memory

of

MICHAEL FARADAY

PREFACE AND INTRODUCTION

In the spring of 1906 I had the honour to deliver before the Royal Institution of Great Britain a lecture "On Recent Progress in Magneto-optics." A distinguished astronomer told me two or three years later that he had read the account of this lecture several times over. I have no doubt that his remark was meant as a compliment, but it signified also a too condensed exposition of the subject. At about the same time came the proposal of the Editor of this series of monographs to supplement my discourse and to give an expansion of it in book form. Consideration of the two communications, both acting in the same direction, easily induced me to prepare the present volume.

A few historical facts in the history of the relations between magnetism and light may be recalled here.

Faraday's fundamental discovery of the magnetic rotation of the plane of polarisation of light propagated parallel to the magnetic field dates from 1845.[1] In 1877 Kerr showed that a magnetisation of an iron mirror modifies the properties of incident plane polarised light. In 1896 and 1897 the present author discovered the resolution of the spectrum lines of a source of light when under the influence of strong magnetic fields.

A double refraction of the plane polarised rays at right angles to the magnetic field was predicted and detected by

[1] Faraday, "Experimental Researches," *Phil. Trans.* 1846, **3**, 1.

Voigt (1898) in incandescent sodium vapour. Cotton and Mouton discovered in 1907 that magnetic fields induce ordinary double refraction in certain pure liquids, an effect intrinsically different from that revealed by Voigt. Jean Becquerel in Paris (1907), J. Becquerel and H. Kamerlingh Onnes in Leyden, and H. du Bois of Holland (in collaboration with G. J. Elias), in Berlin (1908) made admirable experiments at low and very low temperatures on the magnetic resolution of absorption spectra of crystals and salts of the rare earths, of some artificial gems, and of solid solutions of the rare earths in amorphous substances.

It has been justly said that in the investigation of the magnetic resolution of spectrum lines and connected phenomena, theory and experience have been appropriately united. I may refer here in the first place to the intensely stimulating influence of H. A. Lorentz's theories. It is difficult to find adequate words to express my indebtedness to Lorentz's personal inspiration and to his theories. In the present instance, it may perhaps be realised by a perusal of my first publication concerning the chief subject of this volume. I have, therefore, allowed myself to reprint it in Chapter II.

For the later development of my work, I am under the greatest obligations to Prof. W. Voigt, with whom I have been in constant correspondence during many years, and who again showed his sympathy for the present exposition by contributing some recent photographs.

References to the theoretical work of Lorentz and Voigt are therefore distributed throughout the volume.

In the last chapter, some account is given of the behaviour of J. J. Thomson's atom, built up of electrons and of positive electricity, in the magnetic field.

Lately, the view that an atom may be likened to a dissymmetrical magnetic top of the type studied by du Bois has found much favour among physicists. An analogy of this kind was used by Ritz in his remarkable investigations on spectrum series and atomic fields, and it could not remain unnoticed.

Langevin's important paper on the explanation of magnetic and electric double refraction is summarised in the fifth chapter.

The new conception of the Planck energy-element is steadily diffusing into further and further regions of physics. At the present moment, however, Hasenöhrl is, I believe, the only physicist who has made an attempt to connect the distribution of spectrum lines with the energy-elements.

Some results (Chapter V.) concerning the ratio of the number of emitting atoms to the whole number present in a flame invite consideration from the new point of view. As soon as opportunity permits I intend to repeat the quantitative determination in improved circumstances and to make sure of the above-mentioned ratio, tentatively interpreted as meaning that only part of the atoms present are radiating at the very same moment.

The material in the different chapters of the book has been arranged in the main historically. Though many of the observations recorded have now only retrospective interest, at the time they were steps in the new direction opened by the experiments, and this may justify their publication in connected form.

It is now easily realised that the concrete triple event of narrow spectrum lines combined with great resolving power of the analysing contrivance and with strong magnetic field, is decisive for success, if success can be reached, in the magnetic resolution of the lines. The historical development of the subject shows, however, how late these three elements were combined.

According to the plan of this series of monographs, the author of each volume has to bring together the results of his own investigations and in particular directions, and so far as possible to make them intelligible to a wide scientific public. Accordingly, technical details only of use to the specialist have, as a rule, been omitted and the use of symbols restricted to a minimum.

A glance at the bibliography at the end will convince the reader that many authors of important contributions could not even be mentioned in the text.

The rather extensive exposition of my own investigations has somewhat destroyed the balance of the number of pages given to different parts of the subject; especially I regret that, by the limits imposed, only one or two points concerning the above-mentioned phenomena in pleochroic crystals could be touched upon. Most simple and unique are the effects exhibited by the absorption and fluorescence bands of the ruby, which from a magneto-optic point of view may be likened to a crystallised luminuous flame (du Bois). Unique also, but, on the contrary, of wonderful complexity, are the phenomena of resonance and magnetic rotation spectra of metallic and other vapours revealed by Wood's very important investigations. I have avoided dealing with the latter subjects as they have recently been conveniently presented in connected form.[1]

I may be allowed to cite here, with some changes, an opinion expressed in a note to the Paris Academy of Sciences by the late Henri Becquerel and M. Deslandres on April 4th, 1898 : the magnetisation of the spectrum lines has opened a new world of facts which interest the physicist, the astronomer, and even the chemist[2] —especially, I presume, the physical chemists; theirs is much of the future.

[1] R. W. Wood, "Physical Optics." (New York, The Macmillan Co., 1911.)

[2] H. Becquerel et H. Deslandres, "Contribution à l'étude du phénomène de Zeeman," ". . . . l'influence Magnétique . . . . ouvre ainsi un monde nouveau de faits qui intéressent la Physique, la Chimie et même l'Astronomie."

The following list of monographs and general works concerning magneto-optics may be of interest :—

A. Cotton, "Le Phénomène de Zeeman" (Scientia Series), Paris, 1900.

H. A. Lorentz, "Théorie des Phénomènes Magneto-optiques récemment découverts," Rapports présentés au Congrès International de Physique, Paris, 1900, **3**, 1.

C. Runge, "Schwingungen des Lichtes im magnetischen Felde" (H. Kayser, "Handbuch der Spektroskopie," **2**, 611, 1902).

O. J. Lodge. "Electrons." (G. Bell and Sons. London, 1906.)

W. Voigt, "Magneto- und Elektro-optik" (Leipzig, 1908).

H. A. Lorentz, "The Theory of Electrons": Chapter **3**, "Theory of the Zeeman Effect"; Chapter **4**, "Propagation of Light in a Body composed of Molecules"; "Theory of the Inverse Zeeman Effect" (Leipzig, 1909).

H. A. Lorentz, "Theorie der Magneto-optischen Phänomene," Encyclopädie der Mathematischen Wissenschaften, **5**, 3, Heft 2 (Leipzig: Teubner, 1909).

J. Stark, "Prinzipien der Atomdynamik," II Teil. "Die Elementare Strahlung," § 27, Zeeman Effekt, Schwingungszentren der Serienlinien (Leipzig: Hirzel, 1911).

Physicists honour their own craft when endeavouring to picture to themselves the preponderant influence exercised by Faraday's discoveries and conceptions on their science and even on the daily comfort of the human race. To the memory of this sage—the greatest experimental genius the world has produced—the pioneer in magneto-optics as in so many things, I have ventured to dedicate this volume.

I have much pleasure in expressing my heartiest thanks to Miss J. D. van der Waals, Amsterdam, who has made nearly the whole of the translations from the Dutch manuscript. My sincere thanks are due to Prof. A. Fowler, F.R.S., Royal College of Science, South Kensington, for the great care taken in reading the proofs, and for various criticisms. I have also derived great advantage from suggestions made to me by Prof. R. A. Gregory, the Editor of this series of monographs; and I wish to record here my obligations to him for the assistance and courtesy received while the work has been passing through the Press.

CONTENTS

LIST OF PLATES

RESEARCHES IN MAGNETO-OPTICS

CHAPTER I

MODERN SPECTROSCOPES AND RESOLVING POWER

1. The investigations which will be treated in this volume concern a subject in which two departments of physics—light and magnetism—are closely allied. These researches have proved to be particularly adapted to give an insight into the mechanism of emission and absorption of light; they have revealed sharply defined phenomena, which have contributed to the discovery of something about the constitution of the atom. The main fact on which these investigations are founded is the following:—

When a source of light is placed in a magnetic field, the nature of the emitted light undergoes delicate modifications. In the simplest and first observed cases, this change consists in the spectrum lines of the source of light being resolved into two or three components.

The change is exceedingly small even in the strongest magnetic fields; spectroscopes with glass prisms are inadequate to analyse it. It is the high degree of perfection of the modern contrivances of spectroscopy that has rendered it possible to observe the new phenomena and develop our knowledge of them. The special, recent progress of the modern analysing methods and appliances is

B

connected with the names of Rowland, Michelson, Hamy, Fabry and Perot, and Lummer. A short survey of these methods may here be given, at least in so far as they could be used in the investigations under consideration.

When in what follows we refer to wave-lengths, we shall express them in Ångström units (10^{-3} cm.). The two yellow sodium lines indicated as the D_1 and D_2 lines, which can only just be seen separate in a small spectroscope with a glass prism, have wave-lengths of 5896 and 5890 Å.U. The difference of their wave-lengths is therefore 6 Å.U.—about one-thousandth of the whole.

Details at least a hundred times more delicate must be resolved by a spectroscope which is to serve for most of the phenomena to be discussed in this book, and which occur in a region of wave-lengths that never greatly exceeds 1 Å.U.

2. The capacity of a spectroscope to separate vibrations which differ little in wave-length is measured by its separating or *resolving power*.

We owe the theory of the resolving power of gratings and prisms to Lord Rayleigh.[1] This theory, which is as simple as it is beautiful, has given us a means of expressing the value of a spectroscope numerically.

In general, a spectroscope consists of three parts: 1, the collimator, *i.e.*, a linear slit placed in the focus of a lens, so that a parallel beam of light will emerge; 2, the prism, grating, echelon or some other appliance; 3, a lens, in the focal plane of which eye-observations can be made, or photographs obtained. When great resolving power is required, the use of the grating has at present quite superseded the prism.

Fig. 1 represents a greatly enlarged section of a grating. In an opaque screen there are a great number of slits, the

[1] J. W. Strutt, "Investigations in Optics, with Special Reference to the Spectroscope," *Phil. Mag.*, **8**, 261, 403, 477, 1879; **9**, 40, 1880; Lord Rayleigh, Scientific Papers, I., 415.

edges of which are normal to the plane of the figure. The slits are equidistant and have the same width. Let the slit of the collimator be parallel to those of the grating, and the axis of the collimator normal to the plane of the grating. When the slit is illuminated *monochromatically*, a series of images of the shape of the slit is formed by the grating in conjunction with a lens, *M*. One of the images will be at the same place as it would be if the grating were perfectly transparent. The place of this principal image, 0, is independent of the colour.

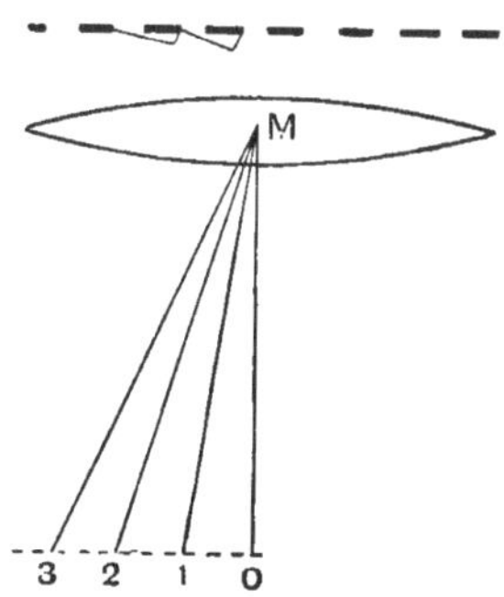

FIG. 1.

On either side of this principal image we find the images formed by diffraction, which are distinguished as images of different "order," and are counted from the principal image.

For violet light the images lie nearer the principal image than for red light. When compound light is used, we get "spectra" always turned with their violet side to the principal image. Even if the slit used is indefinitely narrow, and the light used perfectly homogeneous, the images 0, 1, 2, 3 are of finite width. This is a consequence of the finiteness of the wave-length of light; it introduces a limitation to the power of a spectroscope, as is also the case with microscopes and telescopes.

The distribution of light in the image of a homogeneously illuminated slit is represented in Fig. 2. A central band is bounded by lines where the intensity is zero, and is succeeded by much weaker images. Each of the lines 0, 1, 2 of Fig. 1 exhibits such a distribution of light. If a source of light emits waves of two different wave-lengths, which we denote by λ and $\lambda + d\lambda$, a diffraction image corresponds to each of them. These images are shifted with respect to each other. What the eye observes is the sum of the intensities of the separate images.

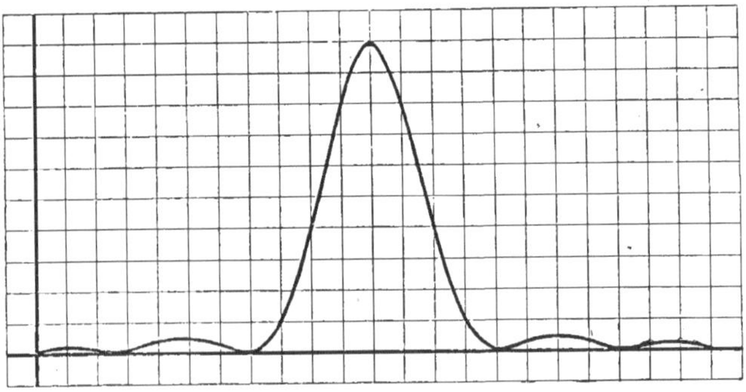

Fig. 2.

If $d\lambda$ is small (Fig. 3), the eye cannot perceive that the light consists of two different wave-lengths. The full line is almost identical with that of Fig. 2.

If the minima of the dotted curves coincide, the central parts would be quite clear of each other, and the double line would be clearly "resolved."

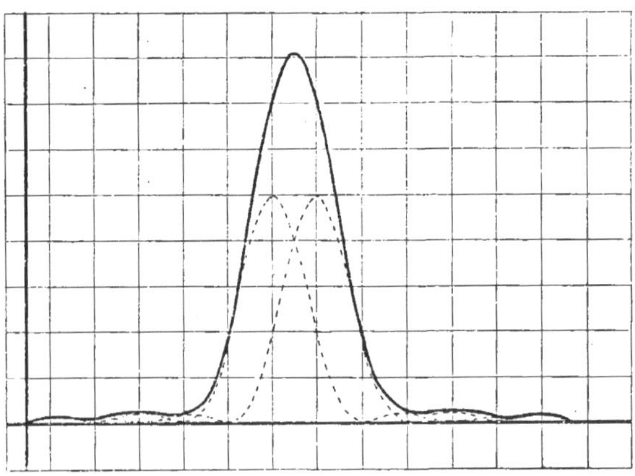

Fig. 3.

Lord Rayleigh found by observation that a line can be just recognised as double if the first minimum of one line coincides with the central band of the other. Then the intensity curve (Fig. 4) exhibits a minimum equal to 0·81 of the maximum value.

When the intensity-curves have the relative position of Fig. 4, we indicate the corresponding difference in wave-

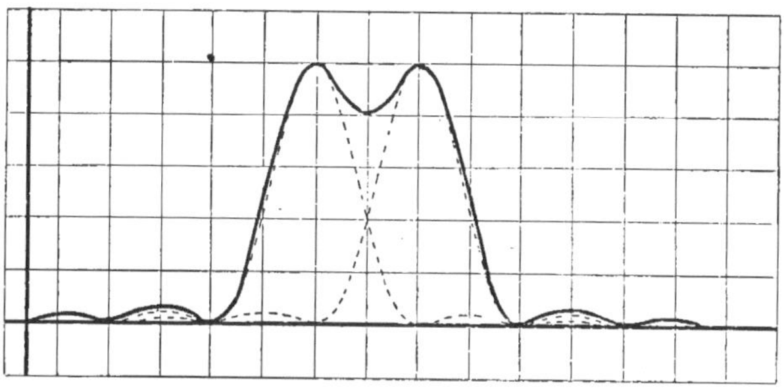

FIG. 4.

length by $\delta\lambda$. Thus lines which differ in wave-length by exactly this amount are just resolved.

The quotient $\frac{\lambda}{\delta\lambda} = R$ may be appropriately called the resolving power.

If n is the *total* number of strips of the grating, m the order of the spectrum considered, Rayleigh proves that

$$\frac{\lambda}{\delta\lambda} = mn = R \quad . \quad . \quad . \quad . \quad . \quad . \quad . \quad (1).$$

This is Lord Rayleigh's fundamental formula for the resolving power of a grating. It shows us that the resolving power depends only on the total number of lines and on the order of the spectrum, but on nothing else. It is quite immaterial how many strips or lines occur in 1 mm.

In the case of the D-lines, $\frac{\lambda}{\delta\lambda}$ is about 1,000 ; so in order to resolve the D-lines a grating with 1,000 lines is required according to formula (1) when the observation is made in the spectrum of the first order; 500 in the second order spectrum, and so on. This simple rule as to the resolving power of a grating has been confirmed by observation.

A spectroscope with a resolving power of 50,000 just permits us to see two lines resolved which are at a distance $\frac{1}{50}$ of that of the D-lines.

3. The *dispersion* of a spectroscope is quite distinct from the resolving power. By dispersion we understand, for a given region of the spectrum, the ratio of the change in the deviation $d\theta$ to the corresponding change in the wave-length $d\lambda$; hence the quotient $\frac{d\theta}{d\lambda}$ expresses the dispersion.

The formula for the dispersion is

$$\frac{mn}{a} = \frac{d\theta}{d\lambda} \quad . \quad . \quad . \quad . \quad . \quad . \quad . \quad . \quad (2),$$

where a is the aperture of the diffracted beam. Theoretically, the dispersion can be made very great, in which case the spectrum becomes very long. It is, however easily possible that, even with a very long spectrum, spectrum lines lying very close together are not separated. For our purpose we want great resolving power, but the dispersion may be quite moderate.

As an analogy illustrating the difference between resolving power and dispersion, we may point out that two representations of the same object may be of the same size, but one may give far more details than the other.

4. Formula (1) is only applicable to a spectroscope with an infinitely narrow slit. This condition cannot be fulfilled in practice. As soon as the slit is widened, the resolving power of the instrument diminishes, whereas the intensity of the light increases. We owe a discussion of the resolving power of a spectroscope with a slit of finite width to Schuster.[1]

Schuster proposes to consider two lines as resolved for wide slits, as also for very narrow slits, when the intensity in the middle of the formed image, which still closely

[1] Schuster, "The Optics of the Spectroscope," *Astrophys. Journ.*, **21**, 197, 1905.

resembles Fig. 4, has a value that amounts to 0·81 of the maximum value.

If the difference of the wave-lengths of the lines which are just resolved with the width of slit used is called $\delta'\lambda$, Schuster defines the "purity" by :—

$$P = \frac{\lambda}{\delta'\lambda} \quad . \quad . \quad . \quad . \quad . \quad . \quad . \quad . \quad . \quad . \quad . \quad (3).$$

With very narrow slits, P becomes the resolving power R. If we put $P = pR$, we may call the p "purity factor." Schuster calculates p for various widths of the slit, and gives in a table the connection between purity and width of slit. He calls the "normal width" of the slit a width equal to $\frac{f\lambda}{4D}$ (where λ is the wave-length, f the focal length of the collimator objective, and D its diameter). The normal width gives a purity which is only 1·4 per cent. less than the maximum possible. With a slit width equal to twice the normal, nearly double the amount of light can be obtained. The purity is 5·7 per cent. less than the maximum. With a slit four times the normal width the intensity of the illumination becomes 3·1, that for normal width being = 1 ; and the loss of purity is 22 per cent. The purity factor becomes 0·495, or about one-half, for a width of slit 7·2 times the normal width, and the intensity is then $\frac{9}{10}$ of the maximum possible, which is four times that obtainable with normal width. A further widening of the slit decreases the purity very rapidly.

The width of the slit must always be rather small. Suppose $f = 200$ cm., $\lambda = 6 \times 10^{-5}$ cm., $D = 10$ cm., then the normal width becomes 0·003 mm. When the slit has a width 7·2 times the normal or 0·022 mm., the resolving power is halved.

The actual width of the slit is of importance in the light of Schuster's discussion ; it is best determined by an optical method. The Fraunhofer diffraction fringes are most suitable for the purpose. With a concave grating, the slit is easily accessible from all sides, and can be left in place.

Monochromatic light is made parallel by means of an auxiliary collimator placed in front of the slit. The parallel diffracted rays are brought to a focus on a photographic plate. After development the distance between the minima can be measured, and from this and the focal length of the lens the slit width is easily calculated.

If the spectroscope has a collimator the method of procedure is exactly similar.

Even if an accurate value of the slit width is not required it is convenient to use the diffraction fringes in order to form an opinion of its order of magnitude. A piece of white pasteboard is simply held at a distance of ten or twenty cm. behind the strongly illuminated slit.

5. Fraunhofer's first gratings consisted of a number of fine metal wires; afterwards he made them by drawing lines on glass plates by means of a diamond. Their resolving power was very slight.

Nobert and Rutherford made greatly improved gratings, with which good results were obtained.

A new epoch for spectrum analysis was, however, opened in 1882 by the work of H. A. Rowland[1] at Johns Hopkins University, Baltimore, U.S.A. Rowland gratings are ruled with a diamond on polished metal mirrors by means of an automatic dividing engine.

To obtain a very high resolving power, very many lines should be traced on the grating, as has been stated already. It is necessary, however, that the lines should be really equidistant. The discrepancy between the theoretical and the practical resolving power is due to want of absolute equality of the spaces between the grooves. The error in the position of the grooves should at most be a very small part of their distance. As Rowland made gratings of a width of several inches with from 10,000 to 20,000 lines to the inch (394 and 788 lines per mm. resp.), the requirements which his dividing engine had to fulfil were very precise. The most essential part of this engine is the screw. Rowland described the method of making this almost perfect screw in the article "Screw" of the

[1] H. A. Rowland, "On Concave Gratings for Optical Purposes," *Phil. Mag.*, **16**, 197, 1883.

9th edition of the "Encyclopedia Britannica." Rowland's engine is still in permanent use in the laboratory of the Johns Hopkins University.[1] In almost all spectroscopic laboratories of the world, whenever great power is required, gratings are used which have been ruled with Rowland's dividing engine.

The largest Rowland gratings have a ruled surface of 5 or 6 inches (by 2 or 3 inches stroke) with 20,000 lines to the inch. According to formula (1) a grating with a total number of about 100,000 lines has in the first order a resolving power of 100,000. In the first order such a grating can separate doublets the distance of which is only $\frac{1}{100}$ of that of the sodium doublet. In the higher orders the resolving power should be increased proportionally to the order under review. The actual resolving power seems, however, never greatly to exceed 100,000. A Rowland grating at the Physical Laboratory of Göttingen University has a resolving power of about 200,000 in the second order.

6. Two lenses are required to obtain sharply defined images with a plane grating. The rays of light must first pass through the lens of the collimator to be made parallel. The grating spreads out the beam of light to a spectrum, which is formed in the focal plane of a second lens. In the Littrow arrangement the same object-glass serves twice.

Rowland's brilliant discovery of the *concave* grating has rendered it possible to observe and photograph the spectra without lenses. These concave gratings are ruled on spherical, concave, reflecting surfaces with a radius of from 3 m. to 6 m., the lines being equally spaced along the chord of the grating. A photograph of a large Rowland grating and grating holder is shown in Fig. 5.

[1] J. S. Ames, "The Present Condition of Rowland's Grating Engine," Johns Hopkins University Circular No. 4, 62, 1906.

For a concave grating we have no longer to take into account the absorption (especially on the ultra-violet) due to the system of lenses, or of the complications, which are inherent to the imperfections of the objectives. The slit,

FIG. 5.

A 6-inch concave grating especially ruled for the author by Prof. Rowland. In the lower part of the figure is seen a lathe-support, a contrivance which is very useful in adjusting delicate optical apparatus.

the concave grating, and the photographic plate form the whole spectrum apparatus. The full length of the spectrum can be photographed with unchanged position of slit and grating, when in the adjustment a rule of Rowland's is observed.

Consider a plane passing through the centre, M, of the spherical grating surface. Let AG be the surface of the grating, the grooves of which are normal to the plane of the diagram (Fig 6).

Describe a circle with AM as diameter. Then, as Rowland proved, an illuminated slit placed at S, somewhere on the circle, and parallel to the grooves of the grating, gives the spectra of different orders all on this circle.

Near M, diametrically opposite the grating, the distances between the spectrum lines are proportional to their wave-lengths. This advantage induced Rowland to use a mounting by which the camera and the grating could be moved with reference to a fixed slit. The greatest stability is, however, secured by a method of mounting

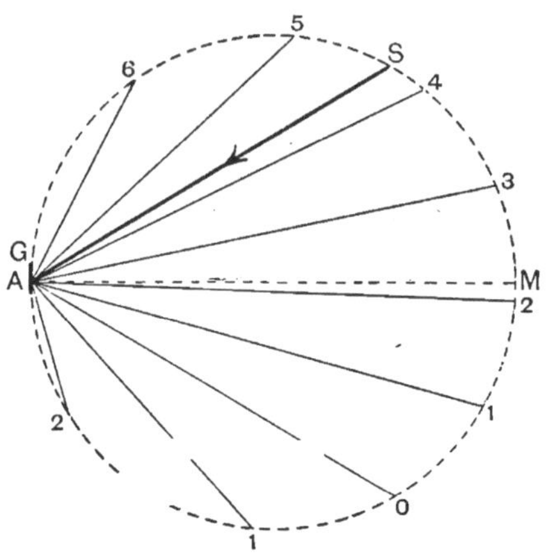

FIG. 6.

with fixed slit and grating, and a camera with photographic plates bent to the radius of curvature of Rowland's circle. The first arrangement of this kind was constructed at Prof. Runge's initiative for the laboratory at Göttingen.

A spectrograph of the same kind at the Amsterdam Laboratory with a 4-inch grating, 52,000 lines and 3 m. radius of curvature, is represented in Fig. 7. On the right at the bottom of the figure is the grating, on the left the slit. The whole apparatus rests in a dark room on two iron beams, which are fastened in the firm walls of the building, very satisfactory stability being thus obtained. Light can be thrown on the grating only through the slit. The source of light is placed in a separate room, which renders it possible to make changes in the light during an experiment.

On the broad brim of the apparatus stands the circular iron camera, consisting of four separate pieces fitting into each other, which are turned in the lathe at the exact radius of curvature, and can be accurately adjusted when

FIG. 7.—THE MOUNTING OF A ROWLAND GRATING.

a spectrum is to be photographed. Photographic plates $7\frac{1}{2}$ cm. long and 4 cm. wide are placed on the camera and fastened with small springs. The division of the camera into four parts and some of the photographic plates in position are visible in the figure.

It is noteworthy that the images formed by the concave gratings are astigmatic. A line of the spectrum image corresponds to a single point of the slit. This astigmatism, generally an advantage, becomes a drawback when the spectra of different parts of a source of light, of which an image is formed on the slit, are to be examined. Spectroscopes with a plane grating are free from this disadvantage. We shall discuss later a method in which a concave grating can be adjusted stigmatically.

Michelson has devised an apparatus with an almost perfect screw, by means of which gratings can be made larger and better even than those of Rowland. In his Nobel lecture Michelson[1] exhibited a grating having a ruled surface nine inches long by four and one half inches stroke (22 × 11 cm.). This grating has 110,000 lines and is nearly perfect in the second order; the resolving power attained is nearly equal to the theoretical 220,000.

The performance of a Michelson grating of 6½-inch ruled surface has been recently tested by Messrs. Gale and Lemon.[2] The test object was the green mercury line, or rather group of lines. The whole group occupies a space about one-fifteenth of that between the sodium lines. Two of the components of the group are at a distance apart of 1/150 of the distance between the sodium lines, and they are so widely separated by the grating that its actual resolving power is from 300,000 to 400,000.

Michelson hopes to attain still greater perfection in the ruling of gratings. There cannot be the slightest doubt as to the supreme importance of these high resolving powers. Several problems demand resolving powers of at least a million, and their solution would open new vistas to spectroscopy and electronic theory.

7. In the *echelon-spectroscope*, Michelson devised an exceedingly ingenious and original kind of grating.[3] It consists of a series of glass plates of perfectly equal thickness, placed on top of each other so as to form steps.

[1] Michelson, "Recent Advances in Spectroscopy," Nobel Lecture. Les Prix Nobel en 1907.

[2] Gale and Lemon, *Astrophys. Journ.*, **31**, 78, 1910.

[3] Michelson, *Journ. de Physique* (3), **8**, 305, 1899; *Astrophys. Journ.*, **8**, 37, 1898.

This system of plates is brought between a collimator and a telescope. The slit of the collimator must be imagined normal to the plane of the drawing. The light is transmitted almost normally. The method of working of the echelon presents close analogy to that of the ordinary grating. The apertures *ab* replace the apertures in the grating (Fig. 8).

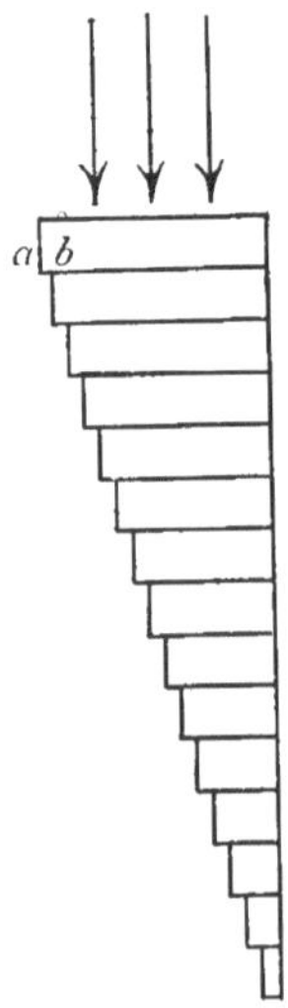

FIG. 8.

While with an ordinary grating the interfering beams of light, on emergence from apertures lying side by side, have only a slight retardation, this becomes very great with the echelon spectroscope. Whereas one beam traverses a glass plate on its way, the other only travels through air, over the same distance. This gives rise to a difference of phase owing to the difference of velocity of propagation in air and glass. If the thickness of the glass plates is 1 cm., the retardation becomes about 10,000 wave-lengths. A great resolving power may be reached with only a quite moderate number of plates, because this is equal, save for a correction, to the product of order and number of apertures, just as for an ordinary grating.

The echelon made by Hilger, which was used in some of my investigations (Fig. 9), contains plates 7·8 mm. thick, the number of effective apertures amounting to 31. The index of refraction of the glass of the echelon is 1·58 for light of the wave-length $\lambda = 5460$ Å.U. (green mercury line), and the resolving power is calculated to be 280,000. An inquiry made expressly to verify this has shown that this theoretical resolving power was actually produced by the instrument.

The result would not have been so favourable if the plates differed in thickness more than 1/20 from a wave-length of light. We cannot help admiring, not only the exceedingly ingenious idea on which the apparatus is based, but also the accuracy to which the art of optics has been carried at present.

It is one of the beautiful qualities of the echelon spectroscope that, in spite of the high resolving power, the intensity of the light of the images remains so great.[1] This is due to the fact that almost all the incident light is concentrated in the direct image and two adjoining spectra of successive order.

On the other hand, only a radiation which is already monochromatic can be investigated with the instrument,

FIG. 9.—THE MICHELSON ECHELON.

because the images of successive orders lie very close together. So a spectrum line or a group of spectrum lines must not be broad, else the images overlap. The portion of the spectrum that can be examined with the above-mentioned echelon may only have a width of at the most 1/10 of the distance of the D-lines.

[1] The repeated reflections between the many surfaces would by loss of light put a rather low limit to the number of available plates. In the echelon made by Hilger this difficulty is overcome by pressing the plates together. The plates are then, remarkably enough, in actual contact, and there is no reflection possible. The introduction of the echelon between collimator and telescope causes a change of focus, which must be corrected, or strange results will be obtained. We owe to Prince Galitzine the best comparison of the theory of the echelon-spectroscope with observation (*Bulletin Acad. imper.*, St. Pétersbourg, (5), **23**, June, Sept., 1905).

In general a preliminary analysis of the light will therefore be necessary. With spectra poor in lines, a small spectroscope, or else a grating, serves for this purpose.

By means of an echelon spectroscope, one can easily ascertain that a number of spectrum lines consist of a main line, accompanied by finer, fainter lines, which are called satellites. This is, among others, the case with many mercury lines. The lines of a vacuum tube charged with cadmium are, on the other hand, very simple. The fact that many spectrum lines are accompanied by satellites was first discovered by Michelson in 1892. His results were obtained by a method which is described in the next paragraph.

8. An interference phenomenon which is closely allied to the famous experiment of "Newton's rings" has found an application in Michelson's *interferometer*. When a beam of monochromatic light is passed through two parallel plates at normal incidence one sees a series of concentric rings alternately bright and dark—a phenomenon to which we shall return later. If the incident light is not really monochromatic, the clearness of the interference phenomenon undergoes alternations when the distance between the plates is changed. The law that governs the change in clearness can be drawn up experimentally, and gives the "visibility-curve," from which, to a certain extent, the composition of the light used can be derived. The genius of Michelson has succeeded in determining with the interferometer the number of wave-lengths of red cadmium light in the standard metre.

Among the further applications of the interferometer made by Michelson we are here particularly concerned with the measurements made by its means at the beginning of the investigations on radiation in magnetic fields.

The echelon spectroscope now enables us directly to effect the fine analysis of the light which at first was only possible by means of the interferometer. For the subject now under review the echelon has superseded the interferometer.

The resolving power of the interferometer is simply equal to the number of wave-lengths contained in twice the distance between the plates.

9. Perot and Fabry's method of the plane parallel lightly-silvered plates excels all other spectroscopic methods by the accuracy with which the theoretical foundations may be practically realised. In its application the principal task of the experimenter has been to effect the perfect parallelism of the two lightly-silvered plates between which the reflection takes place. The principle of the method is as follows:

Light starting from a source falls slightly convergent on two lightly-silvered glass plates with air between, and of

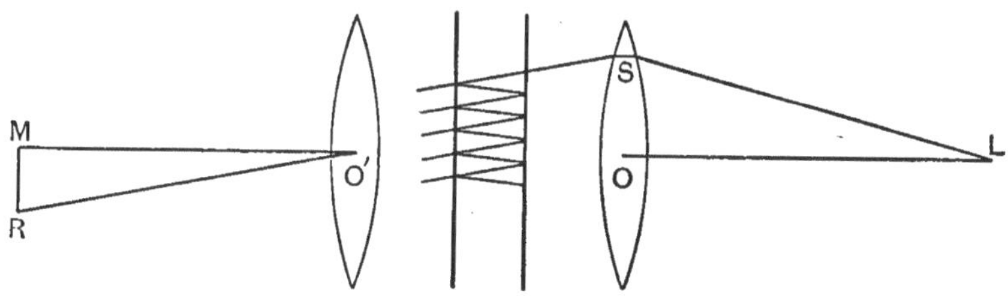

Fig. 10.

which only the inner boundary lines are indicated in Fig. 10. Every incident ray of light gives rise by reflection to numerous others of diminishing intensity.

In the focal plane of a lens these rays can be brought together so as to interfere. According to the laws of optics, a point R in the focal plane of the lens O', corresponds to the parallel incident beams of light. Every ray, LS, yields a point R. When LS is supposed to describe a circular cone, R describes a circle, if the normal on the plates coincides with the axes of the two lenses. The intensity of the light depends on the angle at which the light falls on the plates, but in all the points lying on the same circle the same intensity prevails. In this way it is clear that a system of concentric bright and dark *rings* is observed in the focal plane.

The difference of phase of two successive beams is of the order of magnitude $\frac{2d}{\lambda}$, if d is the distance of the half-silvered plates. This is a high value, even if d be only a few mm.

The resolving power of a grating is determined by the product of the order of the spectrum and the number of apertures. Instead of the number of apertures of Rowland's grating, we have the number of effective issuing beams in Fabry and Perot's apparatus. For well-silvered plates this number may be estimated at 8 or 10[1] for yellow light. We may no doubt safely put the order of magnitude of the resolving power according to the method of parallel semi-silvered plates at $5 \times \frac{2d}{\lambda} = \frac{10d}{\lambda}$.

Hence if $d = 5$ mm., $\lambda = 5000$ Å.U., R becomes $= 5 \times 20{,}000 = 100{,}000$. Accordingly, one can easily make the distance of the plates in Fabry and Perot's apparatus so great that the resolving power of the largest Rowland grating is surpassed.

10. There are two forms in which the method of the parallel plates can be used. The simplest form, which requires less expensive apparatus, has been used by Fabry and Perot, Lord Rayleigh, and Eversheim for the measurement of wave-lengths, and by the author for the investigation of magnetic resolution. This simplest form of apparatus is called *étalon* (Fig. 11). The distance of the silvered plates is here constant. The plates are pressed against rounded distance-pieces by the aid of springs which allow the exertion of a variable pressure.

[1] The reflecting power of silver is smaller for blue than for red light, so that the resolving power is greatest in the red. The semi-transparent silver films may be polished by means of a toilet "powder-puff," after applying to it some "optical" rouge. Pfund, *Astrophys. Journ.*, **28**, 203, 1908.

By change of pressure the steel and the glass can be exceedingly little deformed, and the absolute parallelism of the glass plates is effected, which had already almost been secured by the accurate finish of the distance-pieces.

FIG. 11.—THE FABRY AND PEROT ÉTALON.

Fig. 12 (Plate I) represents an enlarged photograph obtained with Fabry and Perot's apparatus, and may give an idea of the ring system. I am indebted to M. Perot for the original. The distance of the plates amounted to 7 mm.; a vacuum arc tube with mercury was used as source of light. Only the green radiation ($\lambda = 5461$) was used by placing suitable absorbing solutions, which arrest the other kinds of rays, before the source of light. The wires of the micrometer are pointed on the principal line, which is accompanied by satellites.

The different rings are successive spectra of the same line. Difficulties arise from the overlapping of the successive orders of the spectrum if the source of light is not monochromatic. A preliminary analysis of the light makes it possible to overcome these difficulties; but then only one line can be examined at a time. It is

FIG 13.—THE ANALYSIS OF A MAGNETISED SOURCE BY MEANS OF ÉTALON AND SPECTROSCOPE.

better to analyse the light after it has passed through the plate system. The ring system may be projected on the slit of a small spectroscope (Fig. 13), as was done in an investigation by the author in 1907.[1] The ring systems corresponding to the different colours contained in the light are then separated, and a large part of the spectrum can be photographed at once. In Fig. 14 (Plate I) the spectra for the yellow mercury lines ($\lambda = 5791$ and $\lambda = 5770$) partly overlap, the slit being very wide. The ring system on the right corresponds to the green mercury

[1] More recently the firm of A. Hilger, Ltd., has constructed a spectroscope in which the arm carrying the collimator is extended to make room for the Fabry and Perot étalon between the collimator and the prism. This instrument is very compact.

line ($\lambda = 5461$). The faint line on the inside of the rings is a satellite of a wave-length which is $+0{\cdot}085$ Å.U. from the main line. The efficiency of the small apparatus (distance between the plates 5 mm.) is certainly very remarkable.

11. Lummer has devised a plate grating, in which an admirable idea has been realised. I have not had an opportunity as yet to make use of this apparatus in my investigations. Recently I made a few observations with a Lummer-Gehrcke plate; they confirm the high opinion I had formed about the instrument.

12. The simplicity and convenience of manipulation and of calculation of the results is far greater with the diffraction grating than with the newer devices. If, however, extremely high resolving powers are wanted, we must have recourse to these devices, at least at the present moment. But Michelson's endeavours to increase the resolving power of the diffraction grating have met with such a measure of success that there is some hope of still further progress.

All the devices which are now at hand for the student of the finer analysis of the spectrum lines are of comparatively recent date. In 1896, when our investigations on radiation in magnetic fields were begun, the Rowland diffraction grating available was scarcely adequate for the high resolving power required. The radiation phenomena in magnetic fields have shown better perhaps than anything else the advantage of high resolving power,[1] and the want created by these investigations has certainly been conducive to the refined methods of analysis which have been subsequently devised.

One exception must be made to this statement. The Michelson interferometer has a resolving power which is practically unlimited. But this method, which has been

[1] *Cf.* also the opening remarks of Michelson, "Sur le Spectroscope à échelons," *Journ. de Phys.*, [3], **8**, 305, 1899.

applied by Michelson to investigations on the structure of spectrum lines since 1892, requires such a high degree of personal skill that it has scarcely been used outside the Chicago laboratory. I think it necessary to make these remarks for the benefit of the student who has not had an opportunity to become familiar with our subject in its early stages and, looking at the delicate structures, represented, *e.g.*, in our Figs. 33 and 34 (Plate III), would otherwise be at a loss to understand the necessity of so humble a beginning as that recorded in the second chapter.

13. The *electro-magnets* are also of very great importance for the investigation of the magnetic resolution. In 1896 the old type, which had not been changed since Rühmkorff, was still of prevalent use. In 1894 du Bois constructed a large circular electro-magnet according to the new ideas of the magnetic circuit. After this prototype circular magnets were constructed in 1898 which were in many respects more practicable. In my investigations since 1900 I have made use of a semi-circular electro-magnet.

More on the line of Rühmkorff's construction are P. Weiss's electro-magnets of great power, constructed likewise according to the directions of theory.

The highest intensity of the field that can probably be attained in electro-magnets with iron cores[1] has been obtained in the constructions of H. du Bois and P. Weiss. With conical poles having circular end-planes of 4 mm. diameter and a distance of the end-planes of 2 mm. the order of magnitude of the field amounts to 45,000 gauss.

[1] Accurate statements about the attainable field, P. Weiss, *Journ. de Phys.*, **6**, 353, 1907; du Bois, "Neue Halbring Electromagnete"; *Zeitschr. f. Instrumentenkunde*, 376, 1911.

In a few unpublished experiments I have optically investigated an electric spark during its passage between sharply pointed iron (magnet) poles lying very close together. In these experiments the pointed ends are soon worn away by the action of the spark. For some special problems this method deserves application, but the magnetic field is little homogeneous.

A still more limited air-space is impracticable for the experiments we have in view. Moreover, though the distance of the above-mentioned end-planes should be reduced to zero, the intensity of the field would not exceed 55,000.[1] Between the paraboloidal poles of an old-type Rühmkorff magnet a field may be obtained of suitable dimensions of 20,000 gauss.

[1] It might be possible to go much farther by simply using a sufficient solenoid, cooled by liquid air. According to M. Perrin, it only requires a sacrifice of a few million francs to obtain one million gauss in a cubic decimetre (*Bulletin Soc. Franç. de Phys.*, 48, 1907). The project certainly deserves to be brought to the notice of millionaires with philanthropic insight. By means of such a coreless electromagnet our knowledge of matter could be greatly extended.

CHAPTER II

MAGNETIC RESOLUTION OF EMISSION LINES. THE DIRECT EFFECT

14. The wonderful fundamental discovery of a first relation between light and magnetism was made by Faraday in 1845. He succeeded in demonstrating that the plane in which the vibrations of light take place is rotated when the light passes through certain magnetisable bodies along the lines of force. Faraday himself called his discovery, described in the nineteenth series of his "Experimental Researches," "the magnetisation of light and the illumination of magnetic lines of force." His contemporaries did not understand this designation, and perhaps it was more appropriate to what he sought than to what he had found. We shall in this Monograph refer to the magnetic rotation of the plane of polarisation as the Faraday effect. Up to the last years of his life investigations on the relation between light, electricity, and magnetism continued to attract Faraday. The last entry in his Journal, referring to an experiment which is said to have been his last, gives evidence of the delight with which his mind dwelt in a region in which a mutual relation between natural forces might be revealed. For a long time Faraday's discovery remained the sole example of a relation between light and magnetism. In 1877, however, Kerr demonstrated that when polarised light is reflected by magnets a delicate change takes place in the state of the polarisation of the reflected light.

Faraday's and Kerr's discoveries refer to light that is being propagated. We may, however, also inquire into the influence of magnetism on the source of light itself. The idea of the possibility of such a direct action occurred to me while I was occupied with magneto-optical problems, in particular with a quantitative investigation of Kerr's phenomenon in the Leyden Laboratory. In 1896 I succeeded in proving that the emission and absorption of a substance are modified when it is exposed to magnetic forces. This new relation between light and magnetism is, indeed, the most direct and simplest known.

15. The experiments are perhaps best described from the point of view from which they were undertaken. I will therefore allow myself, with some abbreviation, to give the original description, which occurs in my paper, "On the Influence of Magnetism on the Nature of the Light Emitted by a Substance." This paper was laid before the meeting of the Academy of Sciences at Amsterdam on October 31, 1896, by Prof. Kamerlingh Onnes on my behalf.[1] The results of my investigation were communicated to the Academy and appeared in the proceedings of that and the next monthly meeting. I must confess that the exposition is different from that which I should choose now, and I must appeal to the reader's indulgence in this respect.

"On the Influence of Magnetism on the Nature of the Light Emitted by a Substance."

"1. Several years ago, in the course of my measurements concerning the Kerr phenomenon, the thought occurred to me whether the light of a flame, if submitted to the action of magnetism, would perhaps undergo any change. The train of reasoning by which I attempted to illustrate to myself the possibility of this is of minor importance at present; at any rate, I was induced thereby to try the experiment. With an extemporised apparatus the spectrum of a flame, coloured with sodium, placed between the poles of

[1] P. Zeeman, "Over den invloed eener magnetisatie op den aard van het door eën stof uitgestraalde licht," Zittingsverslag, Akademie Amsterdam, **5**, 181, 242, 1896. Translated in *Phil. Mag.*, [5], **43**, 226, 1897.

a Rühmkorff electromagnet, was examined. The result was negative. Probably I should not have tried this experiment again so soon, had not some two years ago the following passage from Maxwell's sketch of Faraday's life come under my notice. Here (Maxwell, 'Collected Works,' II., 790) we read: 'Before we describe this result, we may mention that in 1862 he made the relation between magnetism and light the subject of his very last experimental work. He endeavoured, but in vain, to detect any change in the lines of the spectrum of a flame when the flame was acted on by a powerful magnet.' If a Faraday thought of the possibility of the above-mentioned relation, perhaps it might yet be worth while to try the experiment again with the excellent contrivances of spectroscopy of the present time, as, so far as I know, this has not been done by others as yet. I will take the liberty of briefly stating the results I have obtained up to now.

"2. The electromagnet used was one made by Rühmkorff, and of medium size. The magnetising current, furnished by a storage battery, was in most of the cases 27 amperes, and could be carried up to 35 amperes. The light used was analysed by a Rowland grating, with a radius of curvature of 10 ft., and with 14,438 lines to the inch. The first spectrum was observed with a micrometer eye-piece with a vertical crosswire. An accurately adjustable slit is placed near the source of light under the influence of magnetism.

"3. Between the paraboloidal poles of an electromagnet, the middle part of the flame from a Bunsen burner was placed. A piece of asbestos soaked with common salt was put in the flame in such a manner that the two D-lines were seen as narrow and sharply-defined lines on the dark ground. The distance between the poles was about 7 mm. If the current was put on, the two D-lines were distinctly widened. If the current was cut off, they returned to their original condition. The appearance and disappearance of the widening was simultaneous with the putting on and the breaking off of the current. The experiment could be repeated indefinitely.

"4. The flame of the Bunsen burner was next interchanged with a flame of coal-gas fed with oxygen. In the same manner as before, asbestos soaked with common salt was introduced into the flame. It ascended vertically between the poles. If the current was put on again, the D lines were widened, becoming perhaps three or four times their original width.

"5. With the red lines of lithium, used as carbonate, wholly analogous phenomena were observed.

"6. Possibly the observed phenomenon (§ § 3, 4, 5) will be regarded as not being remarkable at all. One may reason in this manner: widening of the lines of the spectrum of an incandescent vapour is caused by increasing the density of the radiating substance and by increasing the temperature.[1] Now, under the influence of the magnet, the outline of the flame is undoubtedly changed (as is easily seen), hence the temperature and possibly also the density of the vapour is changed. Hence one might be inclined to account in this manner for the phenomenon.

"7. Another experiment is not so easily explained. A tube of porcelain, glazed inside and outside, is placed horizontally between the poles with its axis perpendicular to the line joining the poles. The inner diameter of the

---

[1] Cf., however, also Pringsheim, *Wied. Ann.*, **45**, 457, 1892.

tube is 18 mm., the outer one 22 mm. The length of the tube is 15 cm. The ends of the tube[1] are surrounded by close-fitting pieces, on to which caps provided with parallel glass plates were screwed. The caps are surrounded by little water-jackets. In this manner, by means of a current of water, the copper caps and the glass plates may be kept sufficiently cool while the porcelain tube is rendered incandescent. In the neighbourhood of the glass plates, side tubes provided with taps are fastened to the copper caps. With a large Bunsen burner the tube could be made incandescent over a length of 8 cm. The light of an electric lamp, placed sideways at about two metres from the electromagnet, in order to avoid a disturbing action on the arc, was made to pass through the tube by means of a metallic mirror. The spectrum of the arc was formed by means of the grating. With the eyepiece the D-lines are focussed. This may be done very accurately, as in the centre of the bright D-lines the narrow reversed lines are often seen. Now a piece of sodium was introduced into the tube. The Bunsen flame is ignited and the temperature begins to rise. A coloured vapour soon begins to fill the tube, being at first of a violet, then of a blue and green colour, and at last quite invisible to the naked eye. The absorption soon diminishes as the temperature is increased. The absorption is great only in the neighbourhood of the D-lines. At last the two dark D-lines are visible. At this moment the poles of the electromagnet are pushed close to the tube, their distance now being about 24 mm. The absorption lines are now rather sharp over the greater part of their length. Towards the bottom they increase in width as a consequence of the greater density of the Na-vapour in the lower part of the tube. Immediately after the closing of the current the lines *widen* and are seemingly *blacker*; if the current is cut off, they immediately recover their initial sharpness. The experiment could be repeated several times, till all the sodium had disappeared. The disappearance of the sodium is chiefly to be attributed to the chemical action between it and the glazing of the tube. For further experiments, therefore, unglazed tubes were used.

"8. One may perhaps try to account for the last experiment (§ 7) in this direction: it is true that the tube used was not of the same temperature at the top as at the bottom; further, it appears from the shape of the D-lines (§ 7) that the density of the vapour of sodium is different at different heights. Hence certainly convection currents caused by difference of temperature between the top and bottom were present. Under certain plausible suppositions one may calculate that, by the putting on of the electromagnet, differences of pressure are originated in the tube of the same order of magnitude as those owing to the difference of temperature. Hence the magnetisation will cause, *e.g.*, the denser layer at the bottom to move towards the axis of the tube. The lines become widened; for their width at a given height is chiefly determined by the number of incandescent particles at that height in the direction of the axis of the tube. Although this explanation still leaves some difficulties, there is certainly something to be said in favour of it.

[1] Pringsheim uses similar tubes in his investigation concerning the radiation of gases, *loc. cit.* 430.

"9. The explanation of the widening of the lines attempted in § 8 is no longer applicable to the following variation of the experiment, in which an unglazed tube is used. The inner diameter of the tube was 10 mm., the wall was about 1 mm. thick. The poles of the electromagnet could be moved until the distance was 14 mm. The tube was now heated by means of the blowpipe instead of with the Bunsen burner, and became in the middle part red hot. The blowpipe and the smaller diameter of the tube make it easier to ensure a uniform temperature throughout the tube. This is now higher than before (§ 7), and the sodium lines remain permanently visible.[1] One now can wait till the density of the sodium vapour is the same at various heights. By rotating the tube continuously round its axis I have still further promoted this. The absorption-lines are now equally broad from the top to the bottom. When the electromagnet was put on, the absorption-lines immediately widened along their full width. Now an explanation in the manner of § 8 fails to hold.

"10. I should have liked to study the influence of magnetism on the spectrum of a solid. Oxide of erbium has, as was found by Bunsen and Bahr, the remarkable property of giving by incandescence a spectrum with bright lines. With the dispersion used, however, the edges of these lines were too indistinct to serve my purpose.

"11. The different experiments from §§ 3 to 9 make it more and more probable that the absorption—and hence also the emission-lines of an incandescent vapour—are widened by the action of magnetism. Now if this is really the case, then by the action of magnetism on the free vibrations of the atoms, which are the cause of the line-spectrum, other vibrations of changed period must be superposed. That it is really inevitable to admit this specific action of magnetism will be proved, I think, by the rest of the present paper.

"12. From the conception I had formed of the nature of the forces acting in the magnetic field on the atoms, it seemed to me to follow that with a band-spectrum and with external magnetic forces the phenomenon I had found with a line-spectrum would fail to appear.

"It is, however, very probable that the difference between a band- and a line-spectrum is not of a quantitative, but of a qualitative kind.[2] In the case of a band-spectrum the molecules are complicated, in the case of a line-spectrum the dissociated molecules contain but a few atoms. Further investigation has shown that the representation I had formed of the cause of the widening in the case of a line-spectrum was really true in the main.

"13. A glass tube, closed at both ends by glass plates with parallel faces and containing a piece of iodine, was placed between the poles of the Rühmkorff electro-magnet in the same manner as the tube of porcelain in § 7. A small flame under the tube vaporised the iodine, the violet vapour filling the tube.

"By means of electric light the absorption spectrum could be examined. As the temperature is low, this is the band-spectrum. With the high dispersion used, the bands are resolved into a very great number of fine dark

---

[1] Pringsheim, *loc. cit.*, 456.

[2] Kayser in *Winkelmann's Handbuch*, **2**, 1, 421.

lines. If the current around the magnet is closed, *no* change in the dark lines is observed, which is contrary to the result of the experiments with sodium vapour.

"The absence of the phenomenon in this case supports the explanation, that even in the first experiment with sodium vapour (§ 7) the convection currents had no influence. For in the case now considered, the convection currents originated by magnetism, which I believed to be possible, are apparently insufficient to cause a change of the spectrum ; yet, though I could not observe this by the form of the absorption-line (*cf.* § 7), the band-spectrum is, like the line-spectrum, very sensitive to changes of density and of temperature.

"14. Although the means at my disposal did not enable me to execute more than a preliminary rough measurement, I yet thought it of importance to determine approximately the value of the magnetic change of the period.

"The widening of each of the sodium lines to both sides amounted to about 1/40th of the distance between the D-lines, the intensity of the magnetic field being about 10^4 C.G.S. units. There is therefore a positive and negative magnetic change of 1/40,000th of the period.

"15. The train of reasoning mentioned in (1), by which I was induced to search after an influence of magnetism, was at first the following :—If the hypothesis is true that in a magnetic field a rotatory motion of the ether is going on, the axis of rotation being in the direction of the magnetic forces (Kelvin and Maxwell), and if the radiation of light may be imagined as caused by the motion of the atoms, relative to the centre of mass of the molecule, revolving in all kinds of orbits, suppose for simplicity, circles : then the period, or what comes to the same, the time of describing the circumference of these circles, will be determined by the forces acting between the atoms, and then deviations of the period to both sides will occur through the influence of the perturbing forces between ether and the atoms. The sign of the deviation of course will be determined by the direction of motion, as seen from along the lines of force. The deviation will be the greater the nearer the plane of the circle approximates to a position perpendicular to the lines of force.

"16. Somewhat later I got a clearer insight into the subject by representing to myself the influence exercised on the period of a vibrating system if this is linked together with another in rapid rotatory motion. Lord Kelvin (now 40 years ago[1]) gave the solution of the following problem : Let the two ends of a cord of any length be attached to two points at the ends of a horizontal arm made to rotate round a vertical axis through its middle point at a constant angular velocity, and let a second cord bearing a material point be attached to the middle of the first cord. The motion now is investigated in the case when the point is infinitely little disturbed from its position of equilibrium. With great angular velocity the solution becomes rather simple. Circular vibrations of the point in contrary directions have slightly different periods. If for the double pendulum we substitute a luminiferous atom, and for the rotating arm the rotational motion about the

[1] *Proc. Roy. Soc.*, 1856.

magnetic lines of force, the relation of the mechanical problem to our case will be clear.

"It need not be proved that the above-mentioned considerations are at most of any value as indications of somewhat analogous cases. I communicate them, however, because they were the first inducement to me to undertake my experiments.

"17. A real explanation of the magnetic change of the period seemed to me to follow from Prof. Lorentz's theory.[1]

"In this theory it is assumed that in all bodies small electrically charged particles with a definite mass are present, that all electric phenomena are dependent upon the configuration and motion of these "electrons," and that light vibrations are vibrations of these electrons. Then the charge, configuration, and motion of the electrons completely determine the state of the ether. These electrons, moving in a magnetic field, experience forces which must explain the variation of the period. Prof. Lorentz, to whom I communicated these considerations, at once kindly informed me of the manner in which, according to his theory, the motion of an electron in a magnetic field is to be calculated, and pointed out to me that, if the explanation following from his theory be true, the edges of the lines of the spectrum ought to be circularly polarised. The amount of widening might then be used to determine the ratio between charge and mass, to be attributed in this theory to a particle emitting the vibrations of light.

"The above-mentioned extremely remarkable conclusion of Prof. Lorentz relating to the state of polarisation in the magnetically widened lines I have found to be fully confirmed by experiment (§ 20)."

Paragraphs 18–19 are devoted to a theoretical explanation. In this the term "ions" is everywhere used where we should substitute the word "electrons" for it, as the particles meant by us are now always called, and as has already been done in § 17. Moreover, instead of the method discussed in §§ 18 and 19 of the original paper, we shall use a simplified treatment of Lorentz's theory, dispensing with differential equations.

The electrons in the atoms of a flame may vibrate under the influence of the "elastic" forces which attract them towards the position of equilibrium. Being electrified, they have sufficient hold on the ether to excite electro-magnetic vibrations in it, which, according to Maxwell's theory, constitute light when of

[1] Lorentz. "La Théorie électromagnétique de Maxwell," Leyde, 1892; and "Versuch einer Theorie der electrischen und optischen Erscheinungen in bewegten Korpern," Leiden, 1895.

sufficiently small period. The period of the electrons in the atoms determines the position of the lines in the spectrum; and to every variation of the period of the vibrations corresponds a shifting of the lines in the spectrum.

Let us consider a particle with a single moveable electron, to which, accordingly, a substance corresponds with one spectrum line. An electric particle which moves in a magnetic field experiences an "electro-magnetic" force, which may be found by a simple rule. Let e be the charge of the particle, and v its velocity, the magnetic force being H. Then the value of the force is evH, at least in the simple case which suffices for our purpose—that the velocity is normal to the magnetic force. The direction of the electro-magnetic force is the normal to the plane which may be imagined through the velocity and the magnetic force. For a *positive* particle, the normal may be drawn according to this rule: let the magnetic force ascend, and the velocity be directed towards the south. Then the electro-magnetic force is directed towards the west. If the particle is charged negatively, and if the circumstances are the same for the rest, then the force is directed towards the east. All the motions of the electrons in the atoms of a flame can now be decomposed into three simple motions, chosen so that the influence of the magnetic field can be easily foreseen. The light emitted by the flame is exactly the same as it would be if three groups of electrons were found in it which vibrate in these simple ways.

As a first simple motion we choose a vibration parallel to the lines of force. On the group of electrons which excite this motion, the magnetic force does not exert any influence. The period T remains unchanged.

The other two simple motions are circular, clockwise or anti-clockwise, in planes normal to the lines of force. In a model used in an address delivered before the Royal

Institution the electrons are represented by red balls, the magnetic force by a black arrow (Fig. 15). An electron executing one of these circular motions will, in consequence

Fig. 15.—A Model for the Simple Electronic Motions.

of the magnetic field, be acted on by a force towards the centre or from it, depending on the direction of motion. So the magnetic field will increase or diminish the "elastic" force $f = ar$ by a certain amount, and thus lengthen or shorten the time of revolution. If the motion of the positive electron is right-handed, the electro-magnetic force acts in the same direction as the force f according to the above-stated rule, and the period is shortened (Fig. 16). Hence the period of each of the three motions is T, if the magnetic field is absent, but if the field exerts influence these periods become T, $T + \delta T$, $T - \delta T$, in which δT is a small quantity. Let us now try to find an expression for δT.

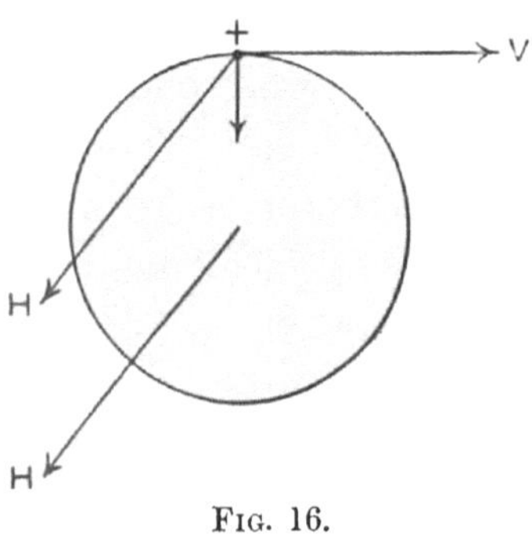

Fig. 16.

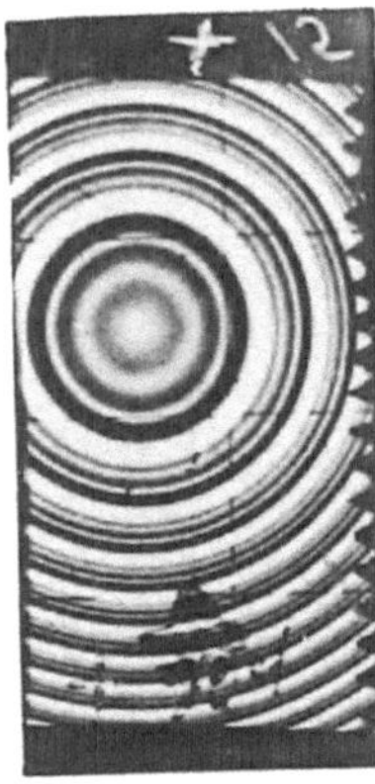

Fig. 12.—Fabry and Perot Rings.

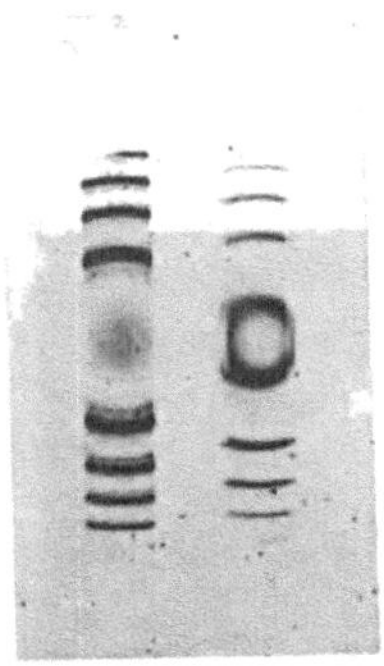

Fig. 14.—Rings for Green and Yellow Mercury Lines.

Fig. 17.

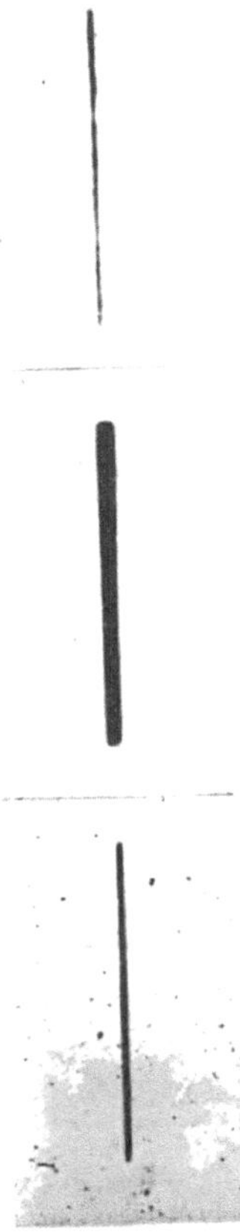

b.

c.

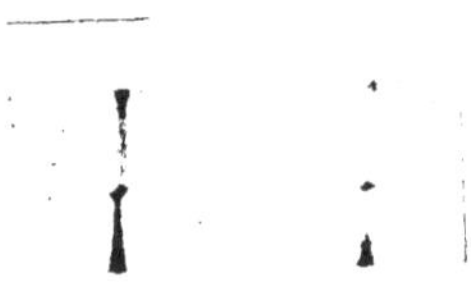

Fig. 18.—Circular Polarisation of Sodium Lines.

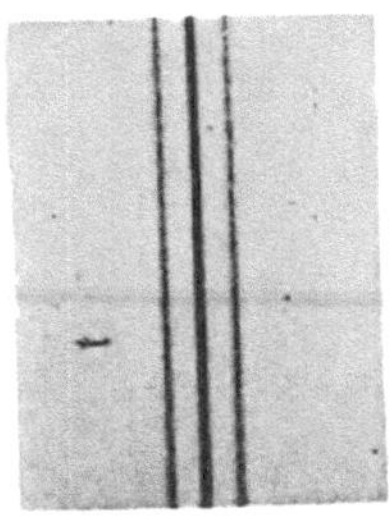

Fig. 19 Zinc 4680.

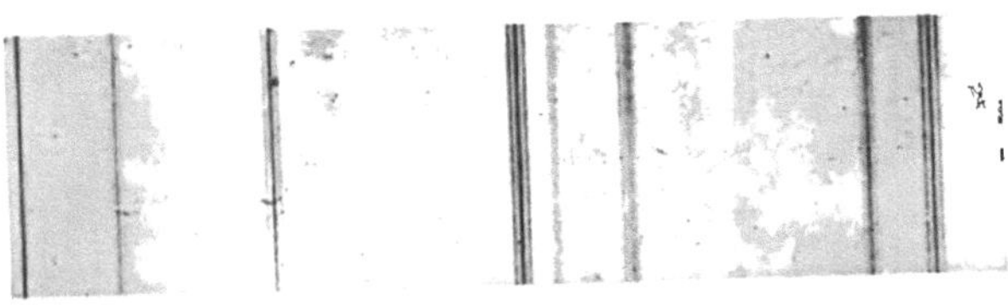

Fig. 20.—Part of Magnetised Iron Spectrum.

Fig. 21.—Linear Polarisation of the two Yellow Mercury Lines.

In the absence of the magnetic force, the angular velocity of the electron is evidently $\omega = \frac{2\pi}{T}$. Let m be the mass of the electron, and let ω be changed into ω', when the magnetic force acts, then

$$a\,r \pm eHv = m\omega'^2 r$$

$$\text{Now} \quad a = m\omega^2 \text{ and } v = \omega' r,$$

$$\text{so} \quad m\omega^2 r \pm eH\omega' r = m\omega'^2 r$$

$$\text{Hence} \quad m(\omega^2 - \omega'^2) = \pm eH\omega'.$$

As ω and ω' differ only little, the whole change being small, it follows from this that:

$$\omega - \omega' = \pm \frac{eH}{2m} \quad . \quad . \quad . \quad . \quad . \quad . \quad . \quad . \quad . \quad (1)$$

Now $1/T$ is the number of revolutions per second. If the variation in this is denoted by $\delta(1/T)$, we may write (1)

$$\delta\left(\frac{1}{T}\right) = \pm \frac{eH}{4\pi m} \quad . \quad . \quad . \quad . \quad . \quad . \quad . \quad . \quad (2)$$

If the change of the period is represented by δT, then $[\delta(1/T)$ being $= -\delta T/T^2]$ the positive or negative change of T is given by:

$$\delta T = \frac{H}{4\pi}\,\frac{e}{m}\,T^2 \quad . \quad . \quad . \quad . \quad . \quad . \quad . \quad . \quad (3),$$

being formula (6) of my original paper. If δT is measured during the experiment, and H and T are known, the ratio e/m of the electron may be determined by the aid of formula (3):

Of course we may also write for (3):

$$\delta\lambda = \frac{H}{4\pi c}\,\frac{e}{m}\,\lambda^2 \quad . \quad . \quad . \quad . \quad . \quad . \quad . \quad . \quad . \quad (4)$$

introducing electromagnetic units, c being the velocity of light.

The disturbance which corresponds to the motion of the electrons can easily be given. First of all the light emitted in a direction *normal* to the lines of force may be examined. Fig. 15 may serve as an illustration here. The vibrations parallel to the lines of force give rise to a ray with electrical vibrations parallel to the lines of force, or a rectilinearly polarised ray of the period T. The light caused by the circular motion is also linearly polarised in the case under consideration, because the orbits are seen, as it were, sideways. Thus we get linearly polarised light with periods $T+\delta T$ and $T-\delta T$, vibrating at right angles to that of the light with the period T. So, all this considered, we shall observe in a spectroscope a three-fold line, a *triplet* with components polarised in the indicated way. The effect now considered may be called the *transversal* effect.

Let us now consider the light which is emitted *parallel* to the direction of the lines of force. We now have to do with the *longitudinal* effect (Fig. 16). It will be clear that the original line has split up into two components with the periods $T+\delta T$ and $T-\delta T$. The light is circularly polarised, that of one component left-handed, that of the other right-handed. Should the resolution into two separate components be impossible on account of the width of the line, or on account of the insufficient intensity of the field, we may yet expect that the light on one edge of the line will have a right-handed circular polarisation, that on the other edge a left-handed.

"20. A confirmation of the last conclusion may be certainly taken as a confirmation of the guiding idea of Prof. Lorentz's theory, To decide this point by experiment, the electromagnet of § 2, but now with pierced poles, was placed so that the axis of the magnet coincided with the line drawn from the eye to the grating. The sodium lines were observed with an eyepiece with a vertical cross-wire. Between the grating and the eyepiece were placed the quarter-wave plate and nicol which I formerly used in my investigation of the light normally reflected from the polished pole of an electromagnet.

"The plate and the nicol were placed relatively in such a manner that right-banded circularly polarised light was quenched. Now according to the preceding, the widened line must at one edge be right-handed circularly polarised, at the other edge left-handed. By a rotation of the analyser over 90° the light that was first extinguished will be transmitted, and *vice versa*. Or, if first the right edge of the line is visible in the apparatus, a reversal of the direction of the current makes the left edge visible. The cross-wire of the eyepiece was set in the bright line. At the reversal of the current the visible line moved! This experiment could be repeated indefinitely.

"21. A small variation of the preceding experiment is the following:—With unchanged position of the quarter-wave plate the analyser is turned round. The widened line is then, during one revolution, twice wide and twice fine.

"22. The electromagnet was turned 90° in a horizontal plane from the position of § 20, the lines of force now being perpendicular to the line joining the slit with the grating. The edges of the widened line now appeared to be plane polarised, at least in so far as the present apparatus permitted to see, the plane of polarisation being perpendicular to the spectral lines. This phenomenon is at once evident from the consideration of § 19. The circular orbits of the electrons, being perpendicular to the lines of force, are now seen on their edges.

"23. The experiments 20 to 22 may be regarded as a proof that the light-vibrations are caused by the motion of electrons, as introduced by Prof. Lorentz in his theory of electricity. From the measured widening (§ 14) by means of equation (3) the ratio e/m may now be deduced. It thus appears that e/m is of the order of magnitude 10^7 electromagnetic *C.G.S.* units. Of course this result is only to be considered as a first approximation.

"24. It may be deduced from the experiment of § 20 whether the positive or the negative electron revolves.

"If the lines of force were running towards the grating, the left-handed circularly-polarised rays appeared to have the smaller period. Hence in connection with § 18 it follows that the negative electrons revolve, or at least describe the greater orbit.

"25. Now that the magnetisation of the lines of the spectrum can be interpreted in the light of the theory of Prof. Lorentz, the further investigation of it becomes specially attractive. A series of further questions already suggest themselves. It seems very promising to investigate the motion of the electrons for various substances, in varying circumstances of temperature and pressure, with varying intensities of the magnetisation. Further inquiry must also decide as to how far the strong magnetic forces which are sometimes supposed to exist at the surface of the sun may change its spectrum.

"The experiments described were made in the Physical Labaratory at Leyden, to the Director of which, Prof. Kamerlingh Onnes, I am under great obligations for his unfailing interest in the present subject."

16. In Fig. 17 (Plate I) some of the phenomena have been represented as they were first observed. Fig. 17, *a* shows the original spectrum line, *b* the magnetically widened line. The linear polarisation of the edges discussed in § 22 of the foregoing extract is shown in *c*, taken in the same circumstances as *b*, with this exception, that a nicol was placed before the slit of the spectroscope which transmitted only horizontal rays.

The following figure (Fig. 18, Plate I) refers to the circular polarisation of the edges for sodium lines (§ 20 of extract, p. 34). In the upper half of the figure both the sodium lines are given when only the right-handed circularly polarised light is transmitted. If the current in the electromagnet is reversed, the lower half is formed. The lines shift their position with reversal of the current in the magnet, which proves that the edges of the lines are circularly polarised in opposite directions.[1]

17. After pure phenomena had been obtained in this way I was convinced that it would be possible to make visible, not only spectrum lines with polarised edges, but also the triplets and doublets predicted by Lorentz's theory.

In communications laid before the Amsterdam Academy of Sciences on May 29, June 26, and October 30, 1897, I was able to state that the observation had succeeded. The investigation is described in a paper entitled: "Doublets and Triplets in the Spectrum Produced by External Magnetic Forces."[2] The observations were rendered possible by the choice of a spectrum line, finer than the sodium lines in a flame, the blue green cadmium line, and by giving the utmost intensity to the field. The 2-inch concave grating used was that of the physical laboratory of Amsterdam, where I had then obtained an appointment; it had a radius of curvature of 1·85 m., 14,438 lines to the inch, with a total number of 27,000 lines. The observations were made in the second order spectrum, which was fairly satisfactory, although not brilliant. An extraordinarily large number of lines, split up into triplets by the magnetic field, have since been

[1] Fig. 18 has been made intentionally under the conditions of the original experiments. The results obtained with stronger fields and higher resolving powers are recorded in § 40.

[2] Proceedings of the Amsterdam Academy and slightly extended, *Phil. Mag.*, [5], **44**, 55, 255, 1897.

observed. As an example I give in Fig. 19 a zinc line ($\lambda = 4580$ Å.U.) and a group of lines from the spectrum of iron (Fig. 20, Plate I). It is interesting to note that the components are so sharp; the effect is not hazy, but very definite.

The linear polarisation of the components may be illustrated by the two yellow mercury lines ($\lambda = 5770$, 5791 Å.U. resp.). (Fig. 21, Plate I.) The distance of the lines is about 21 Å.U., 3·5 times the distance of the two *D*-lines. This spectrum was obtained by means of a spectroscope which gave stigmatic images, so that a point in the image corresponds to a point of the slit. A vacuum tube charged with mercury was used as the source of light. A screen with a small aperture was placed near the vacuum tube. A lens focussed an image of the aperture on the slit of the spectroscope. The size of the aperture in the screen was chosen so that two images of the aperture were formed, one exactly above the other, by a calcspar rhombohedron, adjusted between the lens and the source of light. One image consisted of horizontal, the other of vertical vibrations.

Cornu and König have shown that the circular polarisation may also be made more distinctly visible by the division of the field of vision into two halves, which are oppositely polarised. Then the originally straight spectrum line seems to have acquired a decided break near the line of division between the two fields under the action of the magnetic force. During the photography two quarter wave-plates with horizontal line of demarcation were placed in the beam of light. The main axes in the plates were rotated over a distance of 90° with respect to each other, forming angles of 45° with a horizontal line. After having passed through the plate, an image of which was projected on the slit of the spectroscope, the two opposite circular vibrations were transformed into horizontal and vertical vibrations. With a nicol either one or the other kind of vibration could be quenched.

18. It is very remarkable that the polarisation of the light is so complete. Neither with the linearly polarised components (Fig. 21), nor with the circularly polarised is there to be detected even a trace of the oppositely polarised components. In Chapter VI. we shall discuss the significance of this fact in the light of the general theoretical considerations of Lorentz and Larmor. We shall there also describe the results of a quantitative investigation on the completeness of the circular polarisation. In this investigation a Fresnel rhomb has been substituted for the mica quarter wave-plate. The mica plates, which the investigator can so easily prepare himself, have the drawback that they give the required difference of phase only for a very limited part of the spectrum. As is known, Fresnel's rhomb transforms linearly polarised light vibrating under an azimuth of 45° with the main principal plane of the rhomb, after two total reflections, to circularly polarised light, and this happens *simultaneously for a large part of the spectrum.*

19. Two results given in § 23 and § 24 of my paper (§ 15, p. 35), require some further elucidation. If the lines of force are directed towards us and the motion of the electron is clockwise, then, according to theory, the period of revolution is increased if the electron is negatively charged, but is decreased if the electron is positively charged. It appeared in the experiments that if the lines of force run towards the observer, the right-handed circularly polarised component of the doublet has a greater period than the original line. Thus the electrons causing the radiation of light must be *negatively* charged.

Note on the direction corresponding to the faster propagation in a quarter wave-plate.—The thickness of a plate of mica producing a retardation between the ordinary and the extraordinary rays of a quarter of a wave-length for sodium light is 0·032 mm. Between crossed nicols the quarter wave-plate exhibits the bluish-grey of the first order. The two *principal* directions on the plate parallel to vibrations which are propagated without modification are easily found. For the determination of the direction of revolution in circularly polarised light it is necessary to know which of the two principal directions corresponds to the greater retardation. As is remarked by Wood (*Physical Optics*, 329), this point seems scarcely to be touched upon by text-books.

Wood describes two methods for the purpose. I venture to give here a simpler method I have used for years in my lectures. Place the quarter wave-plate between crossed nicols with its plane perpendicular to the ray of light, and in an azimuth such that the restored light is a maximum. The principal directions on the plate now make angles of 45° with the direction of vibration in the nicols. Then rotate the plate first round one principal direction, then round the other, so that the light traverses a thicker layer. In one case the colour passes from bluish-grey through iron-grey to black ; in the other case through white to yellow and to interference colours of higher order. If the colour becomes yellow one has turned about the direction parallel to the *slower* vibrations. This direction should be marked "slow," the other "fast." The explanation of the method would lead us too far into the optics of crystals. The method is easily extended to other cases, *e.g.*, the determination of the direction of slower vibration in a plate exhibiting the red of the first order.

In the determination of the sign of the circular polarisation it should be borne in mind that the direction of revolution in the polarised light reverses its sign on reflection by a grating, as, indeed, it is in general reversed by a metal mirror. To avoid complications, it is preferable to adjust the analyser before the slit of the spectroscope, if circumstances permit.

20. In the second place, I direct attention to the numerical value found for the ratio of the charge e to the mass m of the electron. In § 23 of the paper of 1896 (§ 15, p. 35) 10^7 electromagnetical units per gramme were found for it as the order of magnitude, and in the paper "over doubletten," &c., of 1897 the more accurate value $1{\cdot}6 \times 10^7$.[1] This is, indeed, the first value found for the exceedingly important quantity e/m in an optical problem. The value found is about 1500 times that of the corresponding value which can be derived for hydrogen from the phenomena of electrolysis.

[1] *Proceedings* of the Amsterdam Academy 1897, and "Doublets and Triplets in the Spectrum produced by External Magnetic Forces," *Phil. Mag.*. [5], **44**, 55, 255, 1897.

This was something entirely new in 1896. After Hertz's famous experiments it had, indeed, been fairly generally accepted[1] that a radiating atom may be considered as an electric oscillator, but our experiments were the first to prove that actually *negative* electricity vibrates in luminous flames, and that a definite, very large, value can be given for the ratio of charge and mass of the vibrating particle. This ratio appeared to be of the same order of magnitude as that which was found at about the same time in the epoch-making investigations of J. J. Thomson in his study of the cathode rays, a department of physics (already associated with the pioneer names of Hittorf, Goldstein, and Sir W. Crookes) which became of increasing importance between the years 1887 and 1895 by the investigations of H. Hertz, Hallwachs, Schuster, Righi, and Lenard,[2] and has since continued to be so.

In our first experiments only flames coloured by metallic salts were examined with regard to their radiation in the magnetic field, but soon after our conclusions were extended to all elements which can be made to radiate in an electric spark or in a vacuum tube.

21. One point should be mentioned here. There may be some doubt left as to the validity of the reasoning which led us to conclude that it is the negative electricity that vibrates in a source of light. Is it not, after all, the same thing whether a negatively charged particle performs a clockwise rotation round a line of force, or a positively charged particle an anti-clockwise rotation in the same orbit? But if so the determination of sign of the moving charge would be illusory.

To this question the following answer may be given.

[1] Cf. H. Ebert, "Elektrische Schwingungen molecularer Gebilde," *Ann. d. Physik*, **49**, 651, 1893; see especially F. Richarz, *Sitz. Ber. Niederrh. Ges. f. Naturk.*, **47**, 113, 1890; **48**, 18, 1891. G. J. Stoney, *Trans.* R. Dublin Soc., **4**, 563, 1891.

[2] Über Kathodenstrahlen, Nobel-vorlesung, 1906.

The movement of the electron performing a clockwise rotation may be replaced by two linear, mutually perpendicular, vibrations. At a distant point somewhere on the axis of the circle, each of these linear vibrations gives rise to an electrical vibration normal to the axis, and parallel to the component which has caused it. There is exactly the same difference of phase of a quarter of a period between the two electric vibrations as exists between the components of the motion of the electrons. Hence the light along the axis will be right-handed circularly polarised. The direction of the motion of the electron and that of the vector which represents the dielectric displacement of the light are the same. Hence if we observe right-handed circularly polarised light, this must be caused by something that performs a clockwise rotation. But if this has once been accepted, then the sign of the charge to be assigned to the moving particle is absolutely determined.

Synthesis of Light.—The negatively charged electron performing a rotation round the lines of force is attracted by a "quasi-elastic" force towards a centre. We may arrange an experiment in which a stream of cathode rays moving with great velocity at right angles to a magnetic field is bent into a *circle.* Of course in this case also electromagnetic waves are emitted. In a field of intensity H, the wave-length λ of the emitted waves is given by $2\pi cm/eH$, c being the velocity of light. Using a vacuum tube with a Wehnelt cathode—a piece of very hot platinum on which there is a speck of lime—large quantities of cathode rays are emitted. If such a tube could be used in a field of $H=10{,}000$ gauss, λ would become $0 \cdot 9$ cm.; for $H=40{,}000$ we would get $\lambda=2$ mm., about the length of the shortest electric waves obtained by von Baeyer. The longest heat-waves isolated by H. Rubens and R. W. Wood have a wave-length of more than one-tenth of a millimetre, and the waves subsequently obtained by

Rubens and von Baeyer are 0·3 mm. in length. The actual performance of the suggested experiment, tried with inadequate means by the author with a vacuum tube in a magnetic field, would give us circular vibrations along the axis of the circle and linear vibrations in its plane. The extremely close analogy between this experiment and the mechanism involved in the studied phenomena of magnetic resolution would then permit us to speak of the synthetic production of ultra-red light.

22. Summarising, we may say that the investigations of 1896 and 1897 described in this chapter have made us acquainted with the following facts:

If a luminous gas which gives one or more sharp lines in the spectrum is immersed in a magnetic field, many lines of the spectrum are resolved into two or three components lying close together. The number depends on the direction in which the light is examined. In a direction parallel to the magnetic lines of force we get *two*, in a direction normal to the magnetic force *three*, components. The lines of the *doublet* are circularly polarised in opposite directions. The lines of the *triplet* are linearly polarised; the light of the middle line is polarised normal to the line of force, the two others in a plane parallel to the line of force. When the lines of force run towards us, the right-handed circularly polarised component has the greater period.

These facts, which were partially predicted by Lorentz's theory, and the significance of which lies in their connection with that theory, are, nevertheless, independent of any theory.

Though partially obtained from experience, our result that e/m for the radiating particles is of the order of magnitude 10^7, and that a negative charge must be attributed to these particles, can only be derived from theoretical considerations, albeit these considerations are now universally accepted.

CHAPTER III

MAGNETIC RESOLUTION OF ABSORPTION LINES. THE INVERSE EFFECT

23. There is an intimate connection between emission and absorption, which in the case of "temperature radiation" is expressed by the law of Kirchhoff; and for cases to which Kirchhoff's developments do not apply there also always exists a parallelism between emission and absorption. Thus to the magnetic resolution of the emission lines there corresponds an entirely analogous change of the absorption lines. The dark lines observed in a continuous spectrum when a beam of white light traverses an absorbing flame are divided and polarised in exactly the same way as was demonstrated for emission lines in Chapter II. This magnetic resolution of absorption lines is often briefly termed the *inverse effect.*

Even in my first experiments, in 1896, I demonstrated this influence of the magnetic field which corresponds to the direct effect. Further experiments have proved that all the peculiarities of the direct effect could also be demonstrated for the inverse effect.

This is a result of great importance. Later we shall see that on the inverse effect very important theoretical considerations are founded, and that it has become of great significance in astrophysics.

24. The arrangement of an experiment for the observation of the inverse effect is very simple. The natural white light of the incandescent positive pole of an arc-lamp traverses a flame which is placed between the poles of an electromagnet, and contains a metallic salt. The issuing light is analysed by means of a spectroscope with a large Rowland grating. Though not for this experiment, it was desirable for experiments which will be described later in this chapter to use a stigmatic spectroscope, so that one point of the image corresponds to one point of the slit. When a plane grating is used, this condition is, of course, satisfied. With a concave grating a particular device has to be applied, as with the usual arrangement one point of the slit corresponds to a line in the image.

In the observation of the inverse effect some peculiarities present themselves, which are best understood if one goes back to the explanation of the inverse effect. This explanation rests on the well-known law of resonance. If in a flame under the influence of a magnetic field there are three periods of free vibrations, then from incident white light vibrations of these three periods will be taken away. The absorption is a selective one, with this peculiarity, that the selection refers, not only to the period, but also to the direction of the vibration. Consider, for example, the central component of a triplet which in the emission spectrum is due to vibrations parallel to the field. From incident white light the only vibrations absorbed correspond as to period as well as to direction of vibration with the middle component. Vibrations perpendicular to the field, though of the period of the unmodified line, pass unimpeded. On the contrary, white light of periods coinciding with those of the outer components is only deprived of its vertical constituents.

It will be clear from this simple consideration what will be observed with natural white light in the above given arrangement of the experiment. When the observation is made at right angles to the lines of force, all the components are observed with the spectroscope, provided the field be strong enough and the vapour has the suitable density. When the observation is made in the direction of the field, only those components are observed which are circularly polarised. All these components are often seen diffuse only, and not black. From the considerations above given, the reason will be clear at once; each of the components absorbs only *half* the incident natural light. With very dilute vapour, no absorption at all, or only very faint traces are seen.

An example of an inverse triplet is furnished by the red Li-line (6708 Å.U.). Fig. 22 (Plate III) is a reproduction of a photograph taken by Prof. Voigt. In the experiment the flame was coloured with a very dilute solution of lithium chloride. The photograph is a positive. What is dark in the photograph is also dark in the field of view during the observation.

25. If the incident beam of light is polarised, the phenomenon becomes entirely different. The absorption lines can then be seen very narrow and black. For definite wave-lengths and modes of vibration, there is now complete absorption.

Let a nicol be placed in the beam of light, and the observation be made at right angles to the horizontal magnetic field. If the nicol is placed so that only vertical vibrations are transmitted, we see the outer components of a triplet black on a bright ground, the middle component is not seen at all, as vertical vibrations which correspond to the original line are not absorbed. If the nicol is turned through an angle of 90°, one sees only the middle line of a triplet, which absorbs horizontal vibrations.

26. To observe along the lines of force, an electromagnet with perforated cores must be used. In the beam of light which passes through the magnetised vapour between the poles, a combination of a nicol and a quarter wave-plate must now be introduced. By this combination the light of the arc-lamp can be circularly polarised in either direction. If right-handed circularly polarised light passes through the vapour, only that component of the doublet is seen black in the spectroscope that emits right-handed circularly polarised light in the direct effect, and *vice versa.*

27. In the preceding sections we supposed that the light that falls on the vapour is polarised. It is, however, of course, also possible to make the light first pass through the vapour, and then observe it by means of a nicol or a quarter wave-plate with a nicol. It is easy to see that it is of no importance to the observations whether the analysis takes place before or after the light has passed through the vapour.

With observations perpendicular to the field, the plane of vibration of the nicol[1] must be vertical if the outer components of the triplet are to be seen black. Horizontal vibrations of the wave-length of the outer components, traversing the flame, are now quenched.

A nicol in the direction of the vibration of light, followed by a quarter wave-plate in a suitable position,[2] converts incident light to right-handed circularly polarised light. If this combination is reversed without change of the relative position, so that now the nicol is preceded by the quarter wave-plate, we have a right-handed circular analyser. It transmits only right-handed circularly polarised light, and no left-handed polarised light. A similar remark holds for left-handed circularly polarised light.

[1] *I.e.*, the plane through the ray and the *short* diagonal of the field of the nicol.

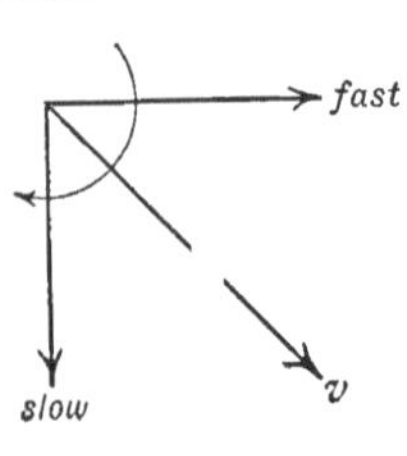

[2] Suppose the direction of vibration of the incident light to be v, making an angle of 45° (as indicated in the figure), with principal section "fast" (see note in § 19 above), then the direction of revolution in the light transmitted by the plate is that shown by the arrow. In mica, which is a negative crystal, the vibrations of the extraordinary rays traverse the plate at the greater velocity.

If in an observation along the lines of force a right-handed circular analyser is used, the absorption line which corresponds to a right-handed circularly polarised emission line, is seen clearly defined and black. In observations of the absorption lines of sun-spots this is the only method of observation possible. All this is very simple; nevertheless, it seems necessary to explain it here, as there has occasionally been some confusion on this subject.

In all the experiments in which a quarter wave-plate is used, a Fresnel rhomb can also be used. This is of advantage when a larger region of the spectrum is investigated.

28. The behaviour of horizontal and vertical vibrations may be studied simultaneously by using a calcspar rhomb according to the suggestion of Cornu and König. By means of it we can obtain two oppositely polarised images of a horizontal slit of suitable width, placed near the magnetic field.

Right-handed and left-handed circular vibrations can be separated on the same plan by the introduction of a Fresnel rhomb between the calcspar and the slit of the spectroscope.

It is, however, of interest to examine also the behaviour of the lines in natural light. A separate examination after the removal of the polarisers might be made. The vapour density ought to be the same in both experiments; but this is difficult to realise.

The desired end is secured more simply and surely, and with only half the labour, by adopting the width of the horizontal slit and the thickness of the calcspar in such a manner that the two images given by the calcspar partially overlap. We now obtain three stripes; the central one exhibits the phenomena as seen without polarising apparatus (see Fig. 23); the upper and lowest

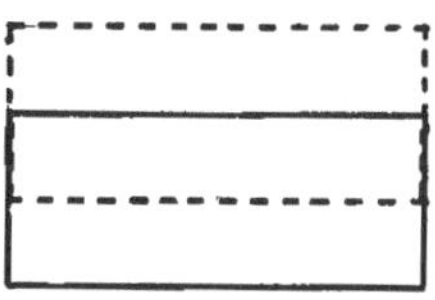

Fig. 23.

stripes show the influence of polarised light on the phenomenon.

By the use of this method all particulars of the phenomenon are simultaneously exhibited; examples of photographs are given later.

29. If the absorption lines are not narrow or if the magnetic field is weak, the components of a magnetically divided line will partially overlap. This partial superposition is the cause of some peculiarities, especially manifest in the inverse effect, and probably also apparent in sun-spot spectra.

The nature of these peculiarities may be illustrated by a few examples. We will consider the case of the magnetic triplet and the magnetic doublet.

In Fig. 24 the ordinates of the curves may represent how much of the light vibrating in the two principal modes corresponding to the three components of the triplet is absorbed if the light is examined at right angles to the lines of force. If natural light passes through a source of light placed in a magnetic field, the total diminution of the light intensity is given by composing the curves.

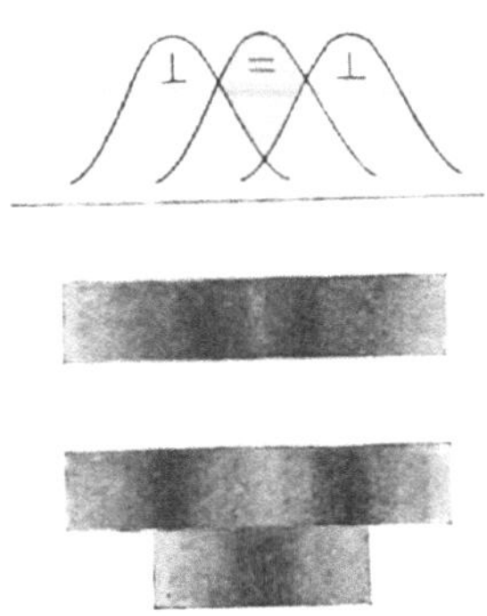

FIG. 24.

If the curves are not separated too far, there must appear in the spectrum *two black bands,* corresponding to the points of intersection of the curves, the resultant curve having then *two maxima.* With further separation of the three curves there are in the resultant curve *two minima* corresponding to the points of intersection. If the three curves are equal, a rule may be easily stated which discriminates between the two cases. We have two maxima in the resulting curve, *i.e.*, two black bands appear in the spectrum, when the point of intersection of each of

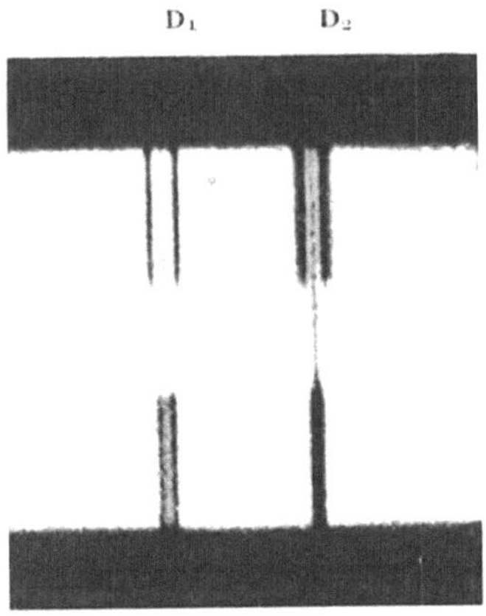

Fig. 27.

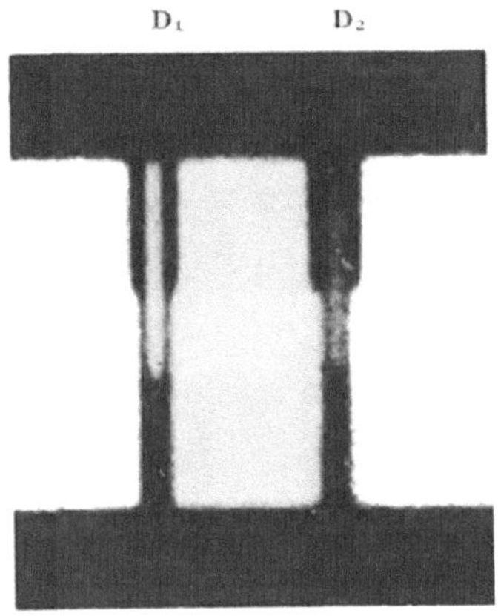

Fig. 29.

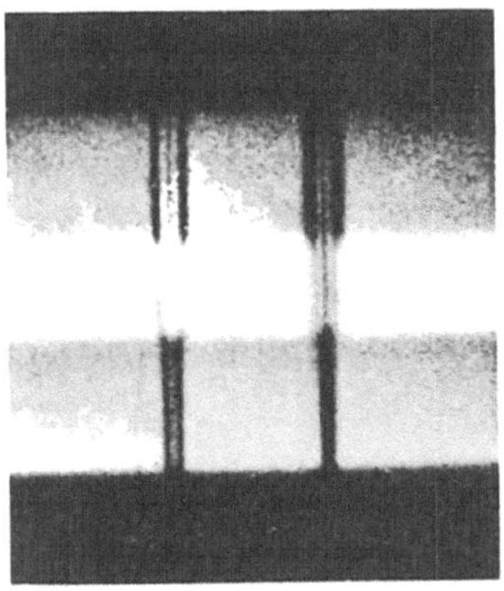

Fig. 28.

Fig. 30

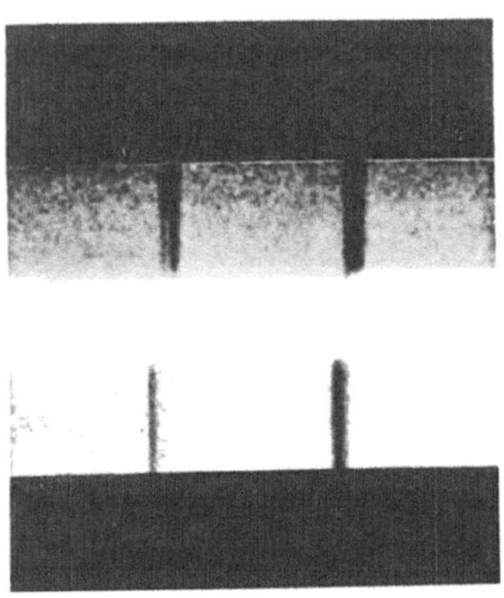

Fig. 31.

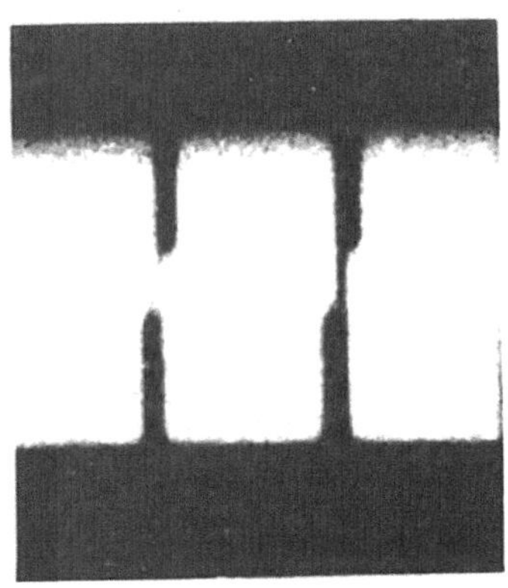

Fig. 32.

the two curves is between the top of a curve and the point of inflection of this curve.

The case of two black bands is represented in the second drawing of Fig. 24.

When a nicol with its plane of vibration vertical is introduced in the beams, two black bands are again seen. The darkest part of these components corresponds to the tops of the curves relating to vertical vibrations. The distance between the components seen with a nicol exceeds that between the black bands first considered. When the plane of vibration of the nicol is horizontal, we see of course only one band.

30. Parallel to the lines of force a partial, not too small, overlapping (cf. § 28) of the components produces a black line with diffuse edges. This case is illustrated diagrammatically in Fig. 25. The two components may be separated by a circular analyser.

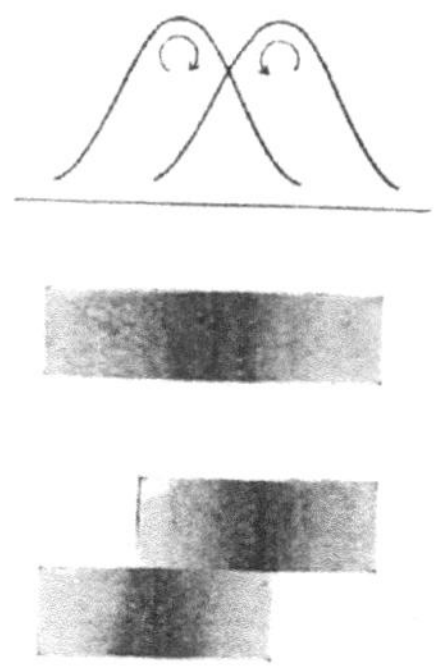

FIG. 25.

These considerations are of some value in the discussion of the magnetic separation in sun-spot spectra (see Chapter VIII.); as a general rule we may expect that the separation of lines in spot spectra becomes more distinct and of larger amount by the use of analysers. The introduction of a nicol into the beams may also reveal lines invisible without analysers.

31. Superposition effects of nearly, though not exactly, the same nature occur if lines with the *same* direction of vibration are superposed and if the continuous source of light emits unpolarised light. The mathematical expression for the propagation of light along the axis of x in an absorbing metallic vapour is

$$e^{-hx} \cos n\left(t - \frac{x}{v}\right),$$

where h is the index of absorption, n the frequency and v the velocity of propagation.

In many cases of propagation of light in a magnetised vapour, the absorption index may also be found by the composition of two curves. For light propagated parallel to the direction of the magnetic field, let the absorption index of right-handed circularly polarised light be h_- and that of left-handed h_+; then if the observation be made transverse to the field, the absorption index of light vibrating perpendicularly to the lines of force becomes [1]

$$h = \tfrac{1}{2}(h_+ + h_-)$$

Hence we may find h simply by composing the lines representing $\frac{1}{2}h_+$ and $\frac{1}{2}h_-$. But of course these h-lines do not *immediately* give the absorption in a layer which is not very thin.

We shall return to the theory of the absorption in a magnetic field in Chapter IX.

32. In 1910 I once more fully investigated the magnetic resolution of absorption lines in collaboration with Winawer.[2] I will select a few examples of the inverse effect of the sodium lines D_1 and D_2 from our paper. These lines, however, are not split up into triplets, but in a less simple way, which will be fully treated in the next chapter. The type and relative amount of the magnetic division of the sodium emission lines D_1 and D_2 are given in Fig. 26.

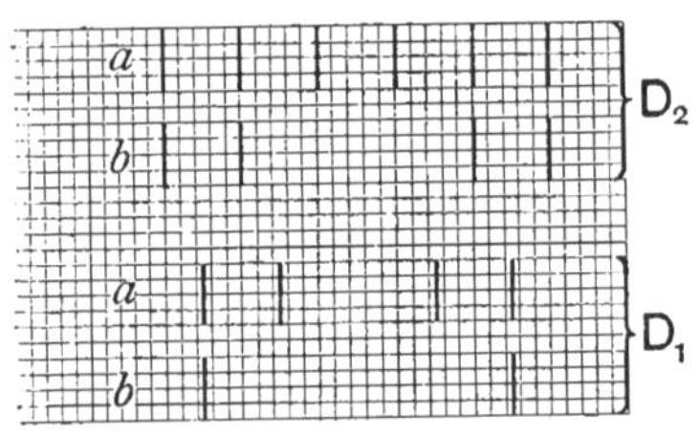

FIG. 26.

a represents the observations when the line of sight is

[1] See Voigt, *Magneto und Elektro-optik*, p. 163, and Lorentz's article in the "Encyclopadie der Math. Wiss.," Band V., 3, Heft. 2, § 50.

[2] P. Zeeman and B. Winawer, "The Magnetic Separation of Absorption Lines in Connection with Sunspot Spectra," *Proc.* Amsterdam Academy, 584, Jan. and Feb., 1910.

at right angles to the magnetic field, *b* when it is parallel to the field. In *a* the two middle components for D_1 and D_2 vibrate parallel to the lines of force; all the other components vibrate at right angles to the lines of force. Along the lines of force circular vibrations correspond to the latter components. In a weak magnetic field D_2 exhibits the triplet type; it is, however, a pseudo triplet; in the direction of the field it is a pseudo doublet. In weak fields, D_1 seems to give doublets in both principal directions.

33. We will now describe some results obtained with the method of observation recorded in § 28 above, when the observation was made in the two principal directions.

Observations perpendicular to the field.—In the upper of the three stripes which are present in the field of view (see § 28), the light vibrates vertically, in the lowest horizontally, whereas the middle part relates to natural light.

Under the influence of the magnetic field we therefore see the vertically vibrating components as narrow black lines. The quartet of the D_1 line, the sextet of the D_2 line, may be seen very clearly by this method. A small disturbance is produced by the narrow reversed lines due to the electric arc light. The intensity of these lines depends upon somewhat variable circumstances of the arc itself. In some cases these lines are almost invisible, in others more prominent. They are to be seen on some of our reproductions (see for the figures mentioned in this and the next paragraph, Plate II); with our present subject they have of course nothing to do.

As regards the central stripe we refer to the remark previously made, that the image of the separation must become, on account of the only partial absorption, rather indefinite and weak. (§ 24.)

The partial superposition of components gives, at least in the case of diluted vapour, the most conspicuous lines. (§§ 29 and 30.) In the case of the quartet, for example, one sometimes sees, instead of four, only two components, situated between the inner and outer ones.

The experiments were made with different vapour densities. The observed phenomena may be classified under three phases:

a. The vapour is *very dilute.* The components are clearly visible in the topmost and lowest stripe. In the central stripe the absorption is either hardly perceptible (Fig. 27) or the components of the quartet and the sextet are seen as separate but weak lines. (Fig. 28.) In this phase of the phenomenon the great difference of definiteness of the central and outer regions is very remarkable. This contrast is still more marked with eye observation. In order to obtain good photographs, it is necessary to increase the density of the vapour above that required for the observation of the very first trace of absorption.

b. Vapour of *intermediate* density. The components in the topmost and lowest stripes are now no longer separately visible or only in the case of the quartet. In the central stripe a superposition of the kind mentioned in § 29 takes place. In place of the quartet an apparent doublet is seen, the components of which are situated between the outer and inner components of the quartet. This case is very clearly represented in Fig. 29. The phenomena exhibited by the sextet (D_2 line) become rather complicated. The superposition phenomenon is often very distinct. The D_2 line on Fig. 29 shows the appearance.

c. With still *denser* vapour, the components become very broad and the magnetic change scarcely visible The *polarisation of the edges* of the broad line may be recognised. This phase is represented in Fig. 30. It corresponds to the emission effect as it was first discovered: a slight change of broad lines in a weak field.

With still greater absorption the influence of the field becomes imperceptible.

All these phases appear with great regularity. If the intensity of the field is known, it seems possible, the resolving power of the spectroscope being given, to deduce the density of the vapour from the nature of the observed phenomena.

34. *Observations parallel to the lines of force.*—In the present experiment the absorbing vapour subjected to magnetic forces is placed between perforated poles. After switching on the current, one sees in the continuous spectrum two dark bands in the case of D_1, four in the case of D_2, as illustrated diagrammatically in Fig. 26. The absorption is now incomplete ; because of some wave-lengths only right-handed circularly polarised light, not left-handed, is absorbed, while the reverse also holds good. In order to observe the separation and the polarisation, a Fresnel rhomb is placed with its principal plane at an azimuth of 45° with the horizon, a horizontal slit being placed in one of the perforated poles. The Fresnel rhomb converts circularly polarised into plane polarised light. By means of a calcspar rhomb three stripes are now obtained. The first phase (very dilute vapour) is represented in Fig. 31.

Vapour of intermediate density (second phase) exhibits the superposition phenomena mentioned in §§ 29 and 30, and diagrammatically illustrated by Fig. 25. In the central strip *one* line, at the position of the unmodified line, is seen surrounded by feebly absorbing regions. Fig. 32 shows these lines for the doublet and the quartet, especially with D_2 ; the effect is very marked. The images given by the grating in Figures 27–32 are enlarged about thirteen times.

CHAPTER IV

COMPLICATED TYPES OF RESOLUTIONS. RELATION BETWEEN RESOLUTION AND SPECTRUM SERIES

35. Very soon after their publication, the experiments described in the second chapter were repeated and extended by a number of investigators. Several new facts came to light, and followed each other in such rapid succession that the same result was repeatedly obtained simultaneously by different physicists. In July, 1897, Michelson communicated the results of an investigation with his interferometer (§ 8). By indirect methods he concluded in fields of 1000–4000 gauss, from the peculiarities of the observed interference phenomenon, that the spectrum lines in the magnetic field split up into two components. For observation in the direction of the lines of force, this is in perfect harmony with the results which I gave in the *Proceedings* of the Amsterdam Academy in the meetings of May and June, 1897, and which have been referred to in the second chapter. The translation of my paper on doublets and triplets is published in the same number of the *Philosophical Magazine* that contains Michelson's communication. In the September number of the same magazine[1] a sequel, dated July 10th, 1897, to the communication on doublets and triplets, was forthcoming. In this second note I was able to account for the difference between my results and

[1] Page 258.

Michelson's, for the case that the observation was made perpendicular to the lines of force. Michelson observed doubling, whereas a triplet was seen by me in this case. I pointed out that the reflection by the mirrors of the interferometer must have a very dissimilar influence on the differently polarised components. This might furnish an explanation of our difference; an explanation which Michelson accepted.

Though I succeeded in making the first *direct* observation of the longitudinal doublet and of the transversal triplet and their polarisations, it is certainly a proof of the efficiency of Michelson's interferometer method that he could ascertain a resolution into two components with very weak fields.

In October, 1897, some photographic representations of the characteristic phenomena which the spectrum lines exhibit in magnetic fields were laid before the meeting of the Amsterdam Academy.

36. In the first half of 1897 my investigations on doublets and triplets were carried out in the physical laboratory of the Amsterdam University, where I was appointed lecturer in January of that year. In consequence of my removal from Leyden and the subsequent change of my work, my experiments advanced but little. I experienced a difficulty of an entirely different nature, viz., the instability of the mounting of the spectrum apparatus in the experiments on doublets and triplets, but this instability made itself felt in a greater degree in the experiments that now seemed of the greatest importance to me.

We have seen how the value of the ratio e/m for the vibrating electrons can be computed from the observed resolution according to Lorentz's elementary theory [see formula (3) p. 33]. The question whether or no this ratio has the same value for different spectrum lines now naturally suggested itself. In the first case, the ratio of the variation of the period to the square of the period $\delta T/T^2$, would have the same value for all lines in the same magnetic field; or in other words [see formula (2) for $\delta(1/T)$, p. 33]: the resolution, measured on the scale of the frequencies, would be the same for all lines in the same magnetic field. I intended particularly to investigate those lines that belong to the same spectrum series.

With a $2\frac{1}{4}$-inch grating of rather feeble intensity having a radius of 6 feet and 14,438 lines to the inch, I began an investigation to which a communication of December, 1897, to the Amsterdam Academy refers. The arrangement was rather unsatisfactory. Grating, slit, and photographic camera were separately mounted on a large wooden table in one of the upper rooms of the laboratory, which were the only rooms at my disposal. The table rested on the wooden floor. A nicol introduced before the slit only admitted the vertical vibrations to the spectroscope. A measurement of negatives obtained with a zinc spark gave a comparatively large resolution, increasing with decreasing wave-length for the lines $\lambda = 4811$, 4722, 4680 Å.U.[1] For the zinc lines $\lambda = 3345$, 3303, 3282, the resolution was too small to admit of measurement. The three first-mentioned lines belong to a so-called second subordinate series, and the three last-mentioned ones to a first subordinate series. So the important conclusions could be drawn from the observations first that e/m is not equal for all lines, and further that different series behave differently in the magnetic field.

[1] In my paper 18·6 : 20·7 : 25·1 has been given for the ratio of the resolution of the mentioned lines. The more recent and accurate measurements show that 18·7 : 21·9 : 25·1 should have been found if the possibility of further resolution of the vertical components is disregarded; just as in almost all the measurements of the time, and even of later times, the intensity of the magnetic field has been over-estimated. Instead of the stated 32,000 gauss, 27,000 gauss would have been nearer the truth.

These stimulating results had been obtained with great difficulty. Every visitor to the upper floor of the laboratory, a movement of the observer who watched the spark, and much more so the permanent traffic in the two adjacent streets, gave rise to almost constant disturbances. Of thirty photographs, at most one could be used. In the course of this part of the investigation it appeared that other observers were in the field, who worked in more favourable circumstances. As there was no hope of obtaining a more stable arrangement for want of funds and room, I was obliged, to my great regret, to abandon for the time being my attempts to photograph the whole spectrum.

Not before 1907 were funds available for the rigid mounting of a 4-inch grating (see § 6). With this whole spectra (iron, nickel, cobalt) have been photographed by my pupils. Even with a time of exposure of about 18 hours, divided over three days, the lines were still perfectly sharp, at least if the temperature outside did not change much.

But in 1897 I was obliged to direct my attention to investigations for which somewhat less stability sufficed. These are investigations on phenomena which accompany magnetic resolution, viz., the magnetic linear and circular double refraction in the neighbourhood of an absorption band. These phenomena will be dealt with in Chapter V. They are comparatively bright, as they appear in the inverse effect. Moreover, they refer only to a small part of the spectrum, so that the focussing of the camera can be renewed now and then, and at first the extreme delicacy of representation which the magnetic resolution demanded was not required. I have entered into these details to make it clear why I could not pursue the promising course which I had at first taken towards the relation between resolution and series of spectra. By this course the desired end could only be reached in favourable circumstances. These favourable circumstances consisted here in great stability and constant temperature of the grating, which were inaccessible to me.

As has been said above (p. 12), there is a Rowland mounting of great stability for a 4-inch grating in the Amsterdam laboratory. But even now (1912) there is no room available where a 6-inch grating could be mounted according to Rowland's method. Our 6-inch concave grating (whole number of lines 55,000, radius 20 ft. 6 ins., 10,000 lines per inch) had to be adjusted *stigmatically* with a view to the experiments which will be described in the following chapters. Then a point in the image corresponds to a point in the slit. The space required for the stigmatic mounting amounts only to half that required for a Rowland mounting (see further, Chapter V.). The "Dutch Society of Sciences" at Haarlem in 1899 enabled me to obtain the before-mentioned 6-inch grating. I wrote to Rowland about the matter, and after about a year I received, to my surprise, the before-mentioned beautiful grating with an inscription which I shall always regard as a particular scientific distinction.

37. In 1897 I succeeded with the small $2\frac{1}{4}$-inch grating, (§ 36) using a current which the Rühmkorff electromagnet can sustain only for a short time, in observing a peculiarity in the resolution of zinc and cadmium lines which proved to be of importance afterwards. Before I had completed and fully discussed my observations, publications appeared by Preston, Cornu, and Michelson, all of them almost simultaneously, on the same phenomenon.

Without a nicol the triplet of the zinc-line 4680 and of the cadmium line 4678 can be easily observed.[1] The lines Cd 4800 and Zn 4722, appeared, however, as groups of four lines.[2] I showed this peculiarity among others to my

[1] If one cadmium and one zinc electrode is used, the two triplets may be observed simultaneously in the field of view.

[2] In a still more delicate analysis it has appeared that the outer components are again split into two. Cd 4800 and Zn 4722 are split up into sextets (see Chapter X. and fig. 72, type *e* and fig. 34, Plate III).

colleagues J. D. van der Waals, Sr., and H. W. Bakhuis Roozeboom. This variation could be explained in my opinion as a reversal. It seemed plausible to suppose that the central part of the broad middle line of the triplet is absorbed in the outer layers of the sparks.

I must point out here that the whole phenomenon was seen on a small scale, and not with the perfection attained afterwards. If so, the notion of "reversal" would have been sooner abandoned.

In the meantime I had also begun measurements with different intensities of the field. There would, no doubt, be no question of reversal if the distance of the components changed proportionally to the magnetic field. The existence of *different* complex resolutions proves that reversal is not adequate to account for all the cases. Reversal, however, is undoubtedly able to exhibit unreal, complex resolutions: "Several iron lines between λ 3700 and λ 3900, which give wide reversals in the arc and spark between iron terminals, can be made to show the Zeeman components also reversed by the use of a strongly condensed spark, so that a triplet appears as a sextuplet." (King.)

There was another circumstance which contributed to my not recognising at once the quartet as a new type of resolution. Greatly impressed by the brilliant results of Lorentz's theory, its probability seemed to me also a proof of its general validity. A simple observation showed that the two middle components of the quadruplet vibrate parallel to the magnetic force. It will certainly be impossible, I considered, that vibrations parallel to the magnetic force can be influenced by it. Yet what seemed improbable proved to be true. A considerable time, however, had to elapse before theory could explain the new type of resolution.

38. For it is actually a type of resolution. I have already said that Preston, Cornu, and Michelson arrived

almost simultaneously at the result that there exist departures from the triplet type. Preston's paper in the *Transactions* of the Royal Dublin Society was published in April, 1898, but dated December 22nd, 1897. Cornu's communication appeared in the *Comptes rendus* of the Paris Academy of Sciences of January 17th, 1898, and Michelson's paper appeared in the February, 1898, number of the *Astrophysical Journal*, and was dated January, 1898.

From all these investigations it appeared that three groups of components occur repeatedly instead of three components. These groups sometimes consist of three, sometimes of two, sometimes of one component. In the last case we get again the original triplet.

Sometimes the middle group consists of two lines, but the outer components remain single; then we have the quartet or quadruplet. To this type belongs the sodium line D_1, the line D_2 exhibiting a sextet in the magnetic field.

Cornu observed with a plane grating mounted according to Littrow's method. Preston's results were obtained with the large grating-mounting of the laboratory at Dublin. Michelson's results were obtained by means of his interferometer method. Some months later Michelson confirmed his results for the greater part by direct observation by means of his newly-invented echelon spectroscope.[1] Thus it appeared, among others, that the green mercury line 5461 is split up in a very complex manner. Under the influence of the magnetic field it becomes a nonet with equidistant components. Michelson gives two more very feeble outer components, but most likely these are not real. The three middle components of the nonet vibrate parallel to the magnetic force. Of the three outer lines on either side, that which is nearest to the original line has the greatest intensity.

[1] Michelson, "The Echelon Spectroscope," *Astrophys. Journ.*, **8**, 37 1898.

We shall give here a few instances of complex resolution. Fig. 33 is a photographic reproduction of the nonet Hg 5461, and Fig. 34 of the sextet Zn 4722 (Plate III). The beauty of these magnetic resolutions is undeniable.

Other new types of resolution were discovered by H. Becquerel and H. Deslandres[1] in 1898, in the course of a fairly extensive investigation of the iron spectrum, with which also a communication by Ames, Earhart and Reese deals.[2] A continuation of this last-mentioned paper is given by the communications of Reese[3] and of Kent.[4]

39. Very important is the relation existing between the resolution of lines which belong to one spectrum series or to corresponding series for different elements. As an introduction to this subject I will state briefly the laws concerning the distribution of the spectrum lines in the absence of a magnetic field. Several spectra contain lines which are closely allied; they form spectrum series.[5] The frequencies of all the lines belonging to a series are expressed by a mathematical formula. Most striking is the existence of connected series for band-spectra. To Deslandres we owe the analysis of the band-spectra, and also the drawing up of formulæ which represent the distribution of the lines in the band-spectra with close approximation. Balmer was the first

[1] H. Becquerel and H. Deslandres, "Contributions à l'étude du phenomène de Zeeman," *Compt. rend.* **126**, 997, 1898; "Observations nouvelles sur le phenomène de Zeeman," *Compt. rend.* **127**, 18, 1898.

[2] J. S. Ames, R. F. Earhart, and H. M. Reese, "Notes on the Zeeman Effect," *Astrophys. Journ.*, **8**, 48, 1898.

[3] H. M. Reese, "An Investigation of the Zeeman Effect," *Astrophys. Journ.*, **12**, 120, 1900.

[4] N. A. Kent, "Notes on the Zeeman Effect," *Astrophys. Journ.*, **13**, 289, 1901.

[5] A summary of our knowledge of the laws of spectrum series of line spectra is given in B. Dunz: "Unsere Kenntnisse von den Seriengesetzen der Linienspectra," Leipzig, 1911. For the whole of the material concerning line spectra, see Kayser's great Handbuch, Vol. V. and Vol. VI. For the important use that can be made of *graphical representations* of the series, compare Lohuizen's Thesis, cited sub [2] (page 62), and *Proc. Amsterdam Academy*, June, 1912.

to give a formula for the simple hydrogen spectrum; not only the four lines of the visible hydrogen spectrum, for which Balmer first drew up his formula, but also the thirty-one lines of the flash spectrum observed during a total eclipse are represented with extraordinary accuracy.

If n represents the reciprocal of the wave-length, or the number of waves in a centimetre, then, according to Balmer, for hydrogen:

$$n = \mathrm{N}_0\left(\tfrac{1}{4} - \frac{1}{m^2}\right) \quad m = 3, 4, 5, 6, \text{ etc.}$$

N_0 being $= 109675$.

If m is increased, the frequency converges to a definite limit. In spectra of several other elements lines also occur which get closer and closer to each other, while they become fainter and fainter. The formulæ proposed for other elements than hydrogen are less simple.

The most important are:

$$n = A - \frac{B}{m^2} + \frac{C}{m^4} \quad \text{(H. Kayser and C. Runge)} \quad . \quad (1)$$

$$n = A - \frac{N_0}{(m+p)^2} \quad \text{(J. R. Rydberg)} \quad . \quad . \quad . \quad . \quad . \quad (2)$$

$$n = A - \frac{N_0}{\left(m + p + \dfrac{q}{m^2}\right)^2} \quad \text{(W. Ritz)} \quad . \quad . \quad . \quad . \quad . \quad (3)$$

$$n = A - \frac{N_0}{\left(m + p + \dfrac{q}{m}\right)^2} \quad \left.\begin{matrix}\text{(E. E. Mogendorff}\\ \text{—W. H. Hicks)}^1\end{matrix}\right\} \quad . \quad . \quad (4)$$

$$n = A - \frac{N_0}{(m+p+qn)^2} \quad . \quad . \quad . \quad . \quad . \quad . \quad . \quad . \quad . \quad . \quad . \quad (5)^2$$

In the formulæ A, p, and q are constants, and m represents the successive whole numbers.

[1] This formula was first given by E. E. Mogendorff, Spectraalreeksen, Thesis for the Doctorate, Amsterdam, 1906. W. H. Hicks, "A Critical Study of Spectral Series," *Phil. Trans.*, London, Ser. A. **210**, 85, 1910.

[2] In T. van Lohuizen, "Bÿdrage tot de kennis van lÿnenspectra," Thesis for the Doctorate, Amsterdam, 1912, the formula (5), which, indeed is one of the forms given by Ritz, is extensively used and an attempt made to classify the series of an element.

Rydberg's formula (2), and also all those following it, contain a universal constant N_0 for all elements. This is very remarkable. The investigations of the formulæ (3)–(5) have proved that for almost all the elements the frequencies of the series lines can be represented by $N_0 = 109675$. We shall hereafter refer to N_0 as the Rydberg constant.

For many elements different series occur. According to Kayser and Runge, they are distinguished as "Principal series" and "first and second subordinate series."[1] The lines of the principal series are exceedingly intense. The first and second subordinate series are also distinguished as the nebulous and the sharp series. In Fig. 35 the

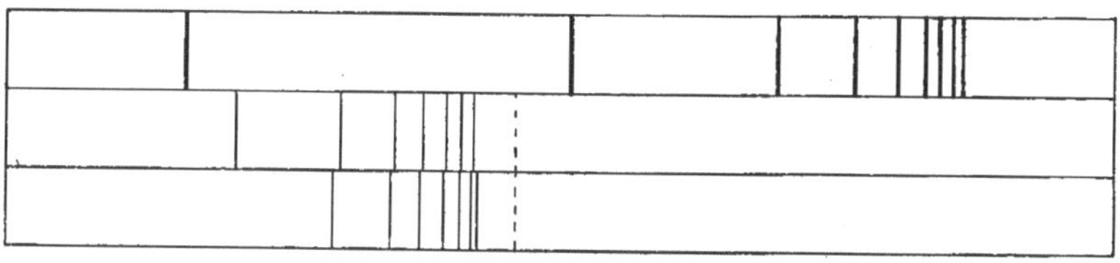

FIG. 35.

position of the series in the scale of the frequencies, or $(1/\lambda)$, is represented diagrammatically. Towards the side of the violet the lines get closer and closer together: they approach the theoretical limit given by $m = \infty$ in the formula. This limit is called by Schuster the root of the series. In the spectra of sodium, potassium, rubidium, and cæsium all the lines are double. Thus there are six series of lines in the sodium spectrum, or strictly speaking eight, Lenard having some years ago found a new series of double lines. Wood recently raised the number of serial double lines observed in the spectrum of sodium, beginning with the D-lines and extending into the remote ultra-violet, to 48. This forms the most

[1] If these series occur in groups of two or three this nomenclature becomes inconvenient. Schuster ("Theory of Optics") suggested naming them respectively the "Trunk Series," the "Side Branch Series," and the "Main Branch Series."

complete series thus far observed. The limit of the first and second subordinate series is the same, and is indicated in the figure by a dotted line. The interval, expressed in frequencies, between the components of the double lines belonging to the first or to the second subordinate series, is independent of the ordinal number m. Hence there are two limits for $m=\infty$ in the subordinate series. For the principal series, on the other hand, there is but one common limit for the frequency of convergence. Consequently, the distance of the double lines of the principal series becomes smaller and smaller as one advances further in the violet.

A remarkable relation between the principal series and the subordinate series was discovered by Rydberg, and independently, but later, by Schuster.

The difference between the convergence frequencies of the principal and the sharp series is equal to the frequency of the first line of the principal series. This law is embodied in Rydberg's formulæ, and is perhaps best remembered in connection with them.

Rydberg's formula for the principal series is

$$\frac{n}{N_0}=\frac{1}{(1+\sigma)^2}-\frac{1}{(m+\mu)^2} \quad . \quad . \quad . \quad . \quad . \quad . \quad . \quad . \quad (6)$$

and for the second subordinate (sharp) series

$$\frac{n_1}{N_0}=\frac{1}{(1+\mu)^2}-\frac{1}{(m+\sigma)^2} \quad . \quad . \quad . \quad . \quad . \quad . \quad . \quad . \quad (7)$$

The difference of the frequencies of the roots $(m=\infty)$ is

$$n\infty - n^1\infty = N_0\left[\frac{1}{(1+\sigma)^2}-\frac{1}{(1+\mu)^2}\right] \quad . \quad . \quad . \quad (8),$$

and this is equal to the value we get if we put $m=1$ in equation (6).

The validity of the law (8) is only approximately verified if Rydberg's equation (2) of the text is used. It becomes much closer if one of the equations (3), (4), or (5), which represent more exactly the different series, is taken.

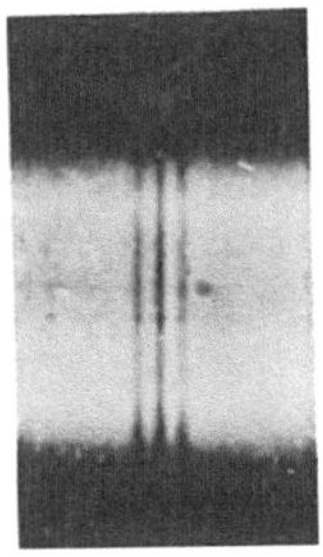

Fig. 22.—Lithium 6708.

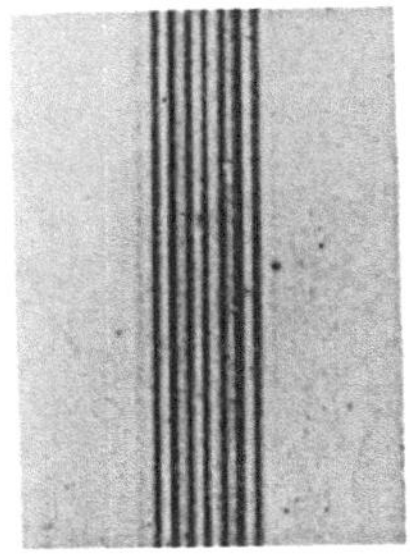

Fig. 33.—Mercury 5461.

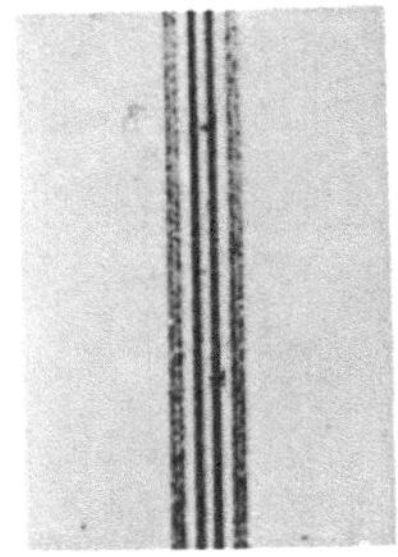

Fig. 34.—Zinc 4722.

Fig. 40.—Magnetic Rotation of plane of Polarisation.

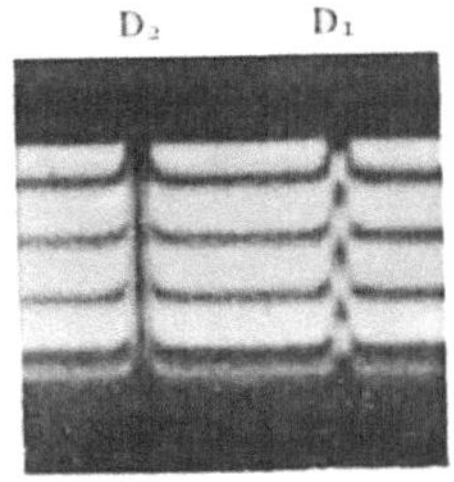

Fig. 41.

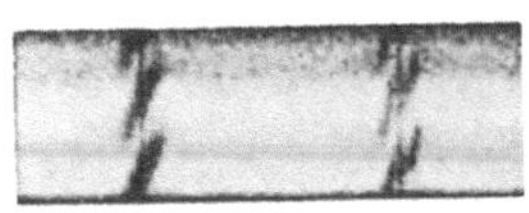

Fig. 42.

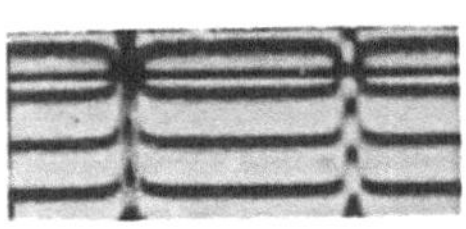

a. 5%. 9100.

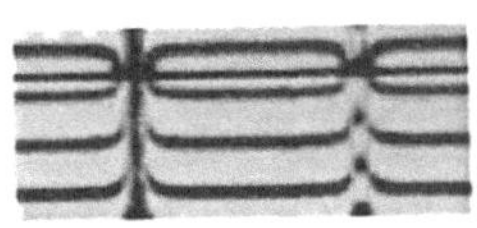

d. 10%. 9100.

b. 5%. 14500.

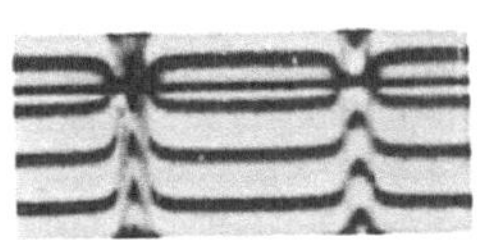

e. 10%. 14500.

c. 5%. 27500.

f. 10%. 27500.

Fig. 43.

From the formulæ by which he represents the series, Rydberg derived a remarkable correspondence between the principal series and the second subordinate series, if they occur in pairs. All the more refrangible lines of the pairs of the principal series are analogous to the less refrangible lines of the sharp series. This correspondence appears also from the intensities. Whereas, *e.g.*, in the principal series the more refrangible line of the pair is the brighter, it is the weaker in the second subordinate series.

In the spectra of Mg, Ca, Sr, Zn, Cd, and Hg series do not occur in pairs, but in triplets.

Series have also been found for other elements. We may mention helium, the spectrum of which can be completely resolved into series. The spectrum consists of six series; among these there are two principal series, two first subordinate series, and two second subordinate series. One of the first subordinate series and one of the second are pair series. The associated principal series also consists of pairs, the first pair having a separation equal to that of the subordinate pairs, but the separation vanishes as the limit of the series is reached.

40. We shall now proceed to the phenomena in the magnetic field.

Th. Preston[1] was the first to state that for an element all the lines belonging to the same series exhibit the same resolution. Not only is the resolution for the different lines of a series similar, but if drawn to the scale of the frequencies it is identical. Also corresponding lines of different elements behave in the same way. Preston gave only few measurements; it is therefore not known to what extent and with what accuracy he investigated the rules given by him. In his publication he only mentioned the lines of magnesium, cadmium, and zinc in so far as they belong to the second subordinate series.

[1] Preston, "Radiating Phenomena in a Strong Magnetic Field," *Phil. Trans.* Royal Soc. Dublin, (2), **7**, 7, 1899.

We owe the most complete and systematic investigation that has appeared on the relation between the magnetic resolution of spectrum lines and the arrangement into series to Runge and Paschen.[1] They begin with an investigation of the radiation of mercury. In the spectrum of this element occur six series: three first subordinate series and three second subordinate series.

The six series exhibit six different types of magnetic resolution. The satellites which accompany the lines of the first subordinate series also present different types. It may be stated with accuracy that the resolutions of the lines conform to the first part of Preston's rule. The three types of the resolutions of the lines of the second subordinate series are given in Fig. 36 *a*. For every series at least two, for many of them three, lines can be investigated.

The second part of Preston's rule refers to the resolution for corresponding series for *different* elements. Observations concerning the second subordinate series of Mg, Sr, Zn, and Cd taught Runge and Paschen that the same resolutions as for Hg occur for corresponding lines of these four elements.

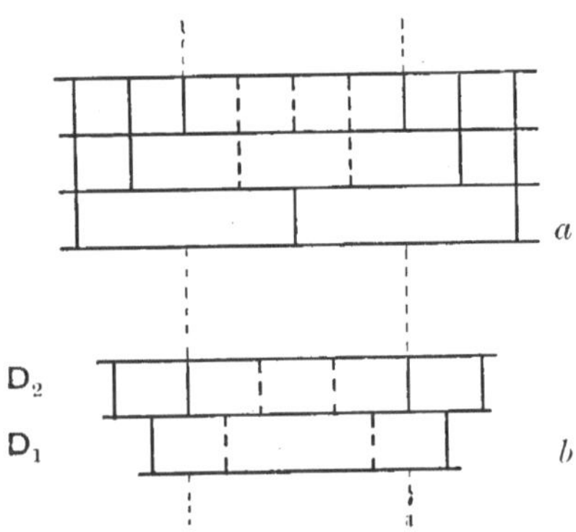

FIG. 36

The resolution of the two sodium lines D_1 and D_2 is represented in Fig. 36 *b*. Series of pairs such as the sodium lines occur also in the spectra of Cu, Ag, Al, Tl, Mg, Ca, Sr, Ba. The behaviour of these doublets in the magnetic field is exactly like that of the sodium lines.

The principal series of copper and silver present the same types of resolution as those of D_1 and D_2. The

[1] C. Runge and F. Paschen, "Uber die Strahlung des Quecksilbers im magnetischen Felde," *Abh. Ak. d. Wiss*, Berlin, 1902, Anhang 1.

second subordinate series exhibits the same types as the principal series, but in reversed order. In each pair of the second subordinate series the line with greater frequency has the type of D_1, that with smaller frequency that of type D_2. This is in harmony with Rydberg's above-mentioned general relation between the two series. (§ 39.) From the magnetic splitting up of the known calcium lines H and K, which exhibit the type of the D-lines, it may be concluded that they belong to a principal series. On the other hand, the Ca-lines 3731·1 and 3706·2 must form a pair in the second subordinate series, for they present the same types in reversed order.

From all these investigations it has appeared that, independent of a spectrum formula, the magnetic resolution proves that the lines of a series are grouped together.

About the series with doublets or triplets we now know from the magnetic resolution that, strictly speaking, we have to do with pairs or trios of series running parallel.

Very remarkable is the behaviour of helium. The lines of this element were ranged into six series by Runge and Paschen. As Lohmann showed, all the lines are split up into similar triplets with the same difference of frequency between their components. Helium is, then, a substance for which for all the lines e/m has the same value, which thus conforms entirely to Lorentz's elementary theory.

For helium, Lohmann's measurements yielded the value $1{\cdot}787 \times 10^7$ for e/m, which is very near Weiss and Cotton's result and that of Gmelin, who found respectively 1·767 and $1{\cdot}771 \times 10^7$ from measurements on the blue zinc line 4680.[1] This is on the supposition that this line had

[1] The concordance of the results of these two beautiful investigations—one performed at Zürich, the other at Tübingen—is certainly very remarkable and gratifying. Gmelin expresses the results in the specific resolution of the *normal triplet* $Z_n = \frac{\Delta\lambda}{\lambda^2 H} \times 10^5$. From Gmelin's measurements follows $Z_n = 9{\cdot}394$, from those of Weiss and Cotton $Z_n = 9{\cdot}375$. $\Delta\lambda$ is the distance between the outer components. For references to the papers of the mentioned authors see pp. 203 and 207 of the Bibliography.

double the separation of the elementary theory (*cf.* Runge's rule, § 41 below).

At the time of these measurements the value of the ratio e/m of the charge to the mass of an electron was smaller by 6 per cent. than the value then accepted. Subsequently the direct measurements of the value of e/m for the Becquerel and cathode rays have shown that it was the value deduced from the magnetic resolution of the spectrum lines that was correct.[1] Indeed recent measurements (Classen, Bücherer, Wolz, Malassez, Bestelmeyer) give as a mean value $1 \cdot 77 \times 10^7$.

Rich material for further investigation has been collected by a number of physicists. Kent, Reese, van Bilderbeek-van Meurs, Miller, Jack, Moore, and King have measured whole spectra. The literature is given in the bibliography at the end of this volume. Some valuable work has been carried out by Purvis concerning many elements not studied by other physicists. Perhaps the most interesting part of Purvis's work is that concerning identical types of resolution in the spectra of the elements gold, lead, tin, antimony, and bismuth, for which no series are as yet known. Purvis has also found some very remarkable states of polarisation, among others for the chromium line 3209·3.

41. The complicated resolutions and their relation with the series fall entirely outside the elementary theory. Some very ingenious hypotheses suggested for their explanation will be discussed in the last chapter, after the phenomena which are related to the inverse effect have been treated; for one of the principal attempts at an explanation of the complex resolutions is in connection with it.

We will here treat one important point. With the complicated resolutions the simple relation furnished by

[1] Cf. A. Cotton, "On the measurements of the Zeeman Effect," *Astrophys. Journ.*, **15**, 213, 1912.

the magnetic resolution into triplets, with the value e/m derived from the cathode rays, seems to have disappeared entirely. A first glance at photographs of complicated resolutions is really discouraging. Fortunately, Runge has made a rule known to us which states that the complicated magnetic resolutions are in simple relation with the normal value of e/m. This rule makes it exceedingly probable that in those cases also the phenomena of light are due to the ordinary negatively charged electron.

To arrive at Runge's rule, we consider the type of resolution of D_2 and D_1 (see Fig. 36). If the halves of the distance of two components are taken as unity, the abscissæ for D_2, reckoned from the middle, are: -5, -3, -1, $+1$, $+3$, $+5$. For D_1 the abscissæ of the components become -4, -2, $+2$, $+4$. All the distances can be expressed by one constant. This is also the case for the three types of the triplet series (Fig. 36*a*). If in this case the distance of two components is called μ, and the distance of the components at the resolution of D_2 $2v$, then the relation $2\mu = 3v$ exists, as appears with great accuracy from the measurements. If e/m is the value of the ratio of charge and mass for the particles of the slow cathode rays, then the *normal* (positive or negative) change a of the frequency of a spectral line in a field H is $\frac{e}{m}\frac{H}{4\pi c} = a = 4{\cdot}692 \times 10^{-5}$ gauss^{-1} cm.$^{-1}$ (*cf.* note, p. 67).

With this constant a the distances of the components, reckoned from the middle, can now be expressed by means of small fractions. This is Runge's rule.[1] For the lines of the type D_1 these distances become $2a/3$, $4a/3$, and for those of the type D_2 $a/3$, $3a/3$, $5a/3$. The distances $a/2$, $2a/2$, $3a/2$, $4a/2$, are found in the second sub-

[1] Runge, "Über die Zerlegung von Spectrallinien im magnetischen Felde," *Physik. Zeitschr.*, **8**, 232, 1907.

ordinate series of mercury, magnesium, calcium, strontium, zinc, cadmium.

Lohmann has found a number of very complicated resolutions in the spectrum of neon with the echelon spectroscope. Runge finds the aliquot parts in this case to be 1/4, 1/5, 1/6, 1/7, 1/11, 1/12.

Runge's rule loses its convincing power as smaller fractions of the interval a, and correspondingly larger multiples, must be chosen in order to represent the measurements.

A rule given by Ritz concerning the commensurability of the resolutions will be discussed in Chapter X.

CHAPTER V

PHENOMENA CLOSELY ALLIED TO THE MAGNETIC RESOLUTION OF ABSORPTION LINES : 1.—MAGNETIC ROTATION OF THE PLANE OF POLARISATION. 2.—MAGNETIC DOUBLE REFRACTION.

42. In order to give an idea of the purpose for which the experiments described in this chapter were undertaken, it is desirable first to summarise the theoretical investigations which inspired them. These experiments refer to the magnetic rotation of the plane of polarisation in metallic vapours, the Faraday effect, and the magnetic double refraction, which can be observed also for metallic vapours in a direction at right angles to the magnetic field. Both phenomena are in close connection with the magnetic resolution of the spectrum lines according to the theory that Voigt has given for it. Voigt starts from the *inverse* effect,—the magnetic resolution of absorption lines treated in Chapter III. Difficulties that confront us in a theory of emission are partly evaded when we consider the absorption. By the introduction into the equations for the propagation of light, of terms which represent a resistance, the absorption can easily be taken into account. By means of the close relation which, according to experience, exists between emission and absorption, peculiarities of the emission can then be discovered.

43. The general train of thought which has led to Voigt's valuable results on the relation between magnetic resolution and the Faraday effect, and between magnetic resolution and magnetic double refraction,[1] can be easily indicated. Some results which have been fully discussed in the chapter on the inverse effect may be recalled. Let a flame which contains metallic vapours absorb vibrations of one definite frequency of incident white light, so that one sharp absorption line is formed in a continuous spectrum. When the flame is placed in a magnetic field, in the simplest case a doublet is observed parallel to the magnetic lines of force. From the incident white light the right-handed circularly polarised constituents, which correspond to the frequency of one component, and the left-handed circularly polarised vibrations, which correspond to the frequency of the other component, are now quite absorbed. We meet with such selective absorptions in some uniaxal pleochroic crystals. Linear vibrations parallel to and normal to the crystallographic axis are absorbed unequally. Unequal absorptions always involve unequal velocities of propagation. An incident wave in a crystal is split up into two waves with unequal velocities of propagation and unequal absorption indices, which are propagated independently of each other.

Still more closely allied to the phenomena in incandescent metallic vapours in a magnetic field are those observed by Cotton[2] in solutions of copper tartrate and chromium tartrate in potash. These coloured liquids exhibit the natural rotation of the plane of polarisation and absorb the two circular rays to an unequal degree.

The thought now naturally suggested itself to Voigt, to whom the physics of crystals owes so much, and who has recently collected his researches on this subject in one

[1] The first-mentioned relation was set forth by Fitzgerald for a particular case at about the same time.

[2] Cotton, *Compt. rend.*, **120**, 989, 1044, 1897.

volume,[1] to trace an analogy between the optical behaviour of pleochroic crystals and sodium vapour in a magnetic field. He was thus led in 1898 to advance the idea that a wave which traverses sodium vapour parallel to the magnetic lines of force is split up into two waves with opposite circular vibrations, which are absorbed in an unequal way, and accordingly will have unequal velocities of propagation. This, however, gives the explanation of the magnetic rotation of the plane of polarisation, for according to the principles of the undulatory theory, the inequality of the velocities of propagation of right-handed and left-handed circularly polarised light follows immediately from the rotation of the plane of polarisation and conversely.

44. For a direction normal to the lines of force, the same considerations may be applied. The analogy with the pleochroic crystals is then complete. In a substance exposed to magnetic forces, different absorption lines correspond to vibrations normal to and parallel to the lines of force; from which we may then conclude to different velocities of propagation for the two vibrations, and hence to the existence of magnetic double refraction.

45. We shall now define the position more closely. It has long been known that the velocity of propagation of light in an absorbing medium, and therefore the index of refraction μ, is a function of the frequency n of the incident light. The same thing applies to the index of absorption.

In the neighbourhood of an absorption band the value of μ is subject to great variations. For very great values of the frequency n, the index of refraction approaches to 1. The function $\mu = f(n)$ is graphically represented in Fig. 37 by the curve $S\ T\ R\ U\ V$. The index rises to a maximum

[1] Voigt, "Lehrbuch der Kristallphysik," Leipsig u. Berlin, Teubner, 1910.

AT, and then falls to a minimum UB. The index of absorption, *i.e.*, the coefficient h, in the expression

$$e^{-hz} \cos n\left(t - \frac{z}{v} + p\right),$$

for the vibration of light, propagated in the direction of the Z-axis, is represented by $M\ L\ N$ as a function of n. The index of absorption has half its maximum value at points which correspond to the maximum AT and the minimum BU. The recent theories of light, both mechanical and electromagnetic, find the explanation of the peculiar course of the index of refraction in the neighbour-

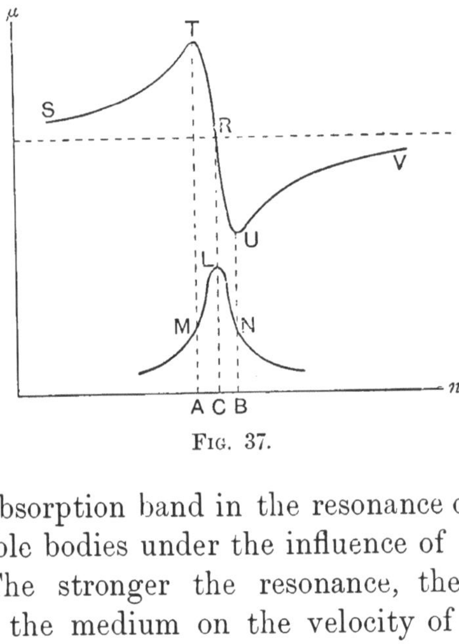

FIG. 37.

hood of an absorption band in the resonance of particles in the ponderable bodies under the influence of incident rays of light. The stronger the resonance, the greater the influence of the medium on the velocity of propagation. The resonance will become greatest when the incident light vibration has a frequency near the frequency of the free vibrations, which is defined by the position of the absorption band. The recent electromagnetic theory of light assumes charged particles in the molecules of the radiating gas instead of the vibrating particles which earlier mechanical theories had to assume. But as we have seen, the charged particles causing the emission are the electrons, and the absorption must therefore also be due to them. Under the influence of the rapidly varying electric forces in the incident beam of light, resonance of the electrons oscillating round a state of equilibrium can then take place in a greater or less degree.

A curve as that of Fig. 37, representing the anomalous dispersion, or as Schuster prefers to say, of the "selective dispersion," can be also experimentally determined for different substances. For sodium vapour, the phenomena of selective dispersion are particularly marked, though the part TU of the dispersion curve could not be examined on account of the exceedingly great absorption *in* the absorption band. There is, however, no doubt about its existence also for sodium vapour. H. Becquerel was the first to examine the refraction of sodium vapour with great dispersion, and particularly also the interesting region between the two D-lines. W. H. Julius and Ebert have studied the subject further in detail, and recently Wood has obtained very important results with sodium vapour in a partial vacuum.

46. In the magnetic field, the phenomena parallel to the magnetic force are the simplest. Voigt has derived from his differential equations for the vibrations of light in the magnetic field the conclusion that the values of the index of refraction μ and of the index of absorption h in the magnetic field are related in a very simple way with the values in the absence of the field. Instead of the absorption curve MLN of Fig. 37, we now get two curves of the same shape, which have been shifted over a distance $eH/2m$ and in opposite directions to their original position. One curve refers to the right-handed, the other to the left-handed ray; if the magnetic force and the propagation of light are both towards us in the line of sight, the curve on the side of the greater frequency must refer to left-handed light (Fig. 38).

This furnishes the explanation of the inverse effect. If

the field be strong enough, we must get two dark components when unpolarised light is sent through the flame. In general, the right-handed and left-handed beams into which the unpolarised light may be thought resolved, will both be absorbed in the way indicated by the curves.

The direction of the displacement is in harmony with what the elementary theory of emission has taught us. The component of the greatest frequency rotates in the direction of the magnetising current.

To each of the two absorption lines corresponds a dis-

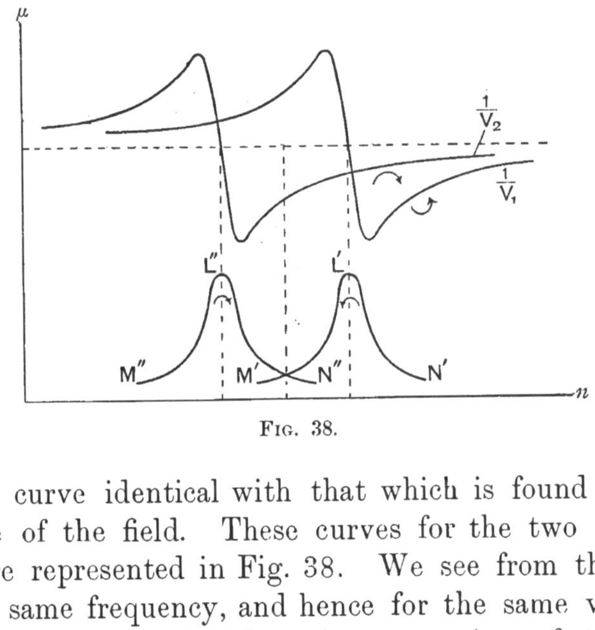

FIG. 38.

persion curve identical with that which is found in the absence of the field. These curves for the two circular rays are represented in Fig. 38. We see from this that for the same frequency, and hence for the same value of the abscissa, the velocities of propagation of the two circular rays are different. In the following section we shall discuss the phenomena caused by this difference of velocities.

47. *Magnetic Rotation of the Plane of Polarisation.*— For the sake of simplicity we shall disregard the difference in the index of absorption of the two rays.

Since the time of Fresnel it has been known that inequality of velocities of propagation of right-handed and left-handed circularly polarised light is equivalent to a rotation of the plane of polarisation of rectilinear vibrations. Let v_1 and v_2 be the velocities of left-handed and right-handed circularly polarised light, then the angle of rotation is, per unit of length,

$$\psi = \frac{1}{2} n \left(\frac{1}{v_2} - \frac{1}{v_1}\right), \quad . \quad . \quad . \quad . \quad . \quad . \quad . \quad . \quad (1)$$

n being the frequency.

When the magnetic force has the direction of the ray of light, ψ is positive if the direction of rotation corresponds

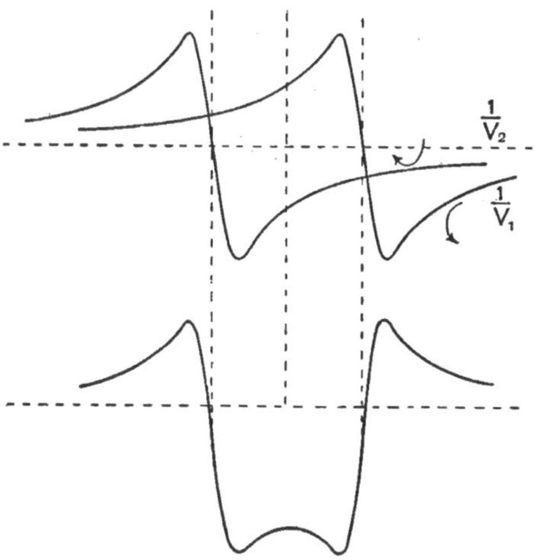

FIG. 39.

to the direction of propagation.[1] This is the case in the graphical representation (Fig. 39). By taking the algebraic difference of corresponding ordinates which represent μ, *i.e.*, $1/v_1$ and $1/v_2$, we at once obtain an idea of the value of ψ which is proportional to it, as appears from the formula (1). It now follows from the construction that the rotation of the plane of polarisation in the neighbourhood of the lines of the doublet must increase rapidly

[1] "The direction of a rotation in a plane and that of a normal to the plane correspond to each other, if an ordinary or right-handed screw, turned in the direction of the rotation, advances in that of the normal." Lorentz, "Theory of Electrons," 3.

and can assume high values. In the region outside the components of the doublet, both on the violet and on the red side, the direction of rotation of the plane of polarisation must correspond to the direction of the current in the electromagnet.[1] On the other hand, in the narrow region between the doublets near the unchanged period, the rotation must become *negative.*

Independently, but simultaneously with the appearance of the theory, Macaluso and Corbino published observations on the magnetic rotation in sodium vapour. These observations referred to the region *outside* the split lines and confirmed the theory in the most brilliant manner. A rotation of about 270° was observed in a sodium flame, which is specifically a much greater rotation than had ever been found in transparent substances. H. Becquerel, Hallo, Wood, Corbino, L. Geiger, and G. J. Elias have further compared the theory with the observations. Wood's work, particularly, on the subject contains an admirable comparison with theory. The maximum rotation observed by him extends to seven complete revolutions. Some results of Hallo's work, which was made in the Amsterdam Laboratory, will be described in Section 51.

In the experiment of Macaluso and Corbino, the light, after having traversed the analyser, is analysed with a spectroscope of fairly large resolving power.

We owe the simplest experiment in which the strong rotation of the plane of polarisation plays the principal part to Righi,[2] who published it some time before the publication of Macaluso and Corbino's paper. But the inverse effect for light parallel to the lines of force is also

[1] It is interesting to note that for *transparent* substances an equation between rotation and dispersion has been established. Using this equation and the value of the magnetic effect given by the elementary theory, Siertsema, from his observations on the magnetic rotation of different gases, deduces values of e/m which are in rather satisfactory agreement with the standard value.

[2] Righi, *Compt. rend.*, **127**, 216, 1898 ; *ibid.* **128**, 45, 1899.

a factor. A horizontal ray of white light traverses the field of an electromagnet with pierced poles parallel to the lines of force. Crossed nicols are placed before and behind the instrument, as in Faraday's experiment. As the sodium flame in the field emits two kinds of circularly polarised rays, it absorbs these same radiations, but does not stop the radiations polarised in the opposite directions. These remaining circularly polarised rays cannot be extinguished by a nicol.

This experiment can be made before a large audience.[1] A brilliant yellow spot appears on a screen upon which an image of the circular hole of the magnet is focussed. This strong light undoubtedly is due to the Faraday rotation near the absorption lines.

48. All these observations refer to the region *outside* the split components. The negative rotation between the components of the doublet required by theory was observed in the experiments which I made at Voigt's suggestion in 1902,[2] and concerning which some particulars may here be given.

In the observations on rotation in sodium vapour, Fresnel's system of quartz wedges was used—a combination of two wedges rotating in opposite direction to the plane of polarisation.

The axes of the quartz crystals are at right angles to the bounding planes of the combination. The light of an arc lamp polarised by a nicol first traverses the magnetic field with the flame in the direction of the lines of force, then the Fresnel prism, which is placed very close to the field, then a lens and an analyser. Now we see a system of sharp bands on the slit. The central band is uncoloured; the others are coloured. In a stigmatic spectroscope we observe a system of nearly horizontal dark bands on

[1] Cf. also Zeeman, *Proc.* Royal Institution, March 30th, 1906.

[2] P. Zeeman, "Observations on the Magnetic Rotation of the Plane of Polarisation in the Interior of an Absorption Band," *Proc. Acad. Amsterdam*, May, 1902.

the spectrum. These correspond to that place in the slit where the direction of vibration in the incident light is at right angles to the direction of vibration of the analyser. If the direction of vibration in the light that falls on the quartz prism is rotated, the place of the dark bands will change. If this change is different for light of different frequency the bands will be deformed. The amount of the displacement is proportional to the amount of the rotation, and it can be easily expressed in absolute measure, a displacement equal to the distance of the bands corresponding to a rotation of 180°. The whole course of the rotation with the wave-length is seen at a glance by means of the method described. The reproduction of a photograph (Fig. 40, Plate III) from the Amsterdam laboratory depicts the horizontal, dark interference bands, which are greatly deformed in the neighbourhood of the thick absorption lines of sodium under the influence of the magnetic force. On either side of the absorption lines there is a positive, large rotation of the plane of polarisation, as Macaluso and Corbino were the first to find. The region between the absorption lines is under the influence of two lines, which causes the rotations at the right and at the left of the lines to be not quite the same.[1]

49. Following my original description,[2] I will now describe the manner in which the interference fringes are transformed on changing the quantity of sodium vapour, the field being kept constant.

The following phenomena were observed for D_1:

If the quantity of sodium in the magnetic field was extremely small, the interference fringe exhibited at the place of the reversed sodium line a protuberance—let us say *downward*—the lines of the doublet being somewhat stronger just above the interference fringe.

[1] The method of interference bands was also applied in 1901 by Corbino (*Atti*, Acc. Lincei, **10**, 137), but only to dense vapours.

[2] Zeeman, *loc. cit.*, note 2, p. 79.

Fig. 45.—Magnetic double Refraction near Sodium Lines.

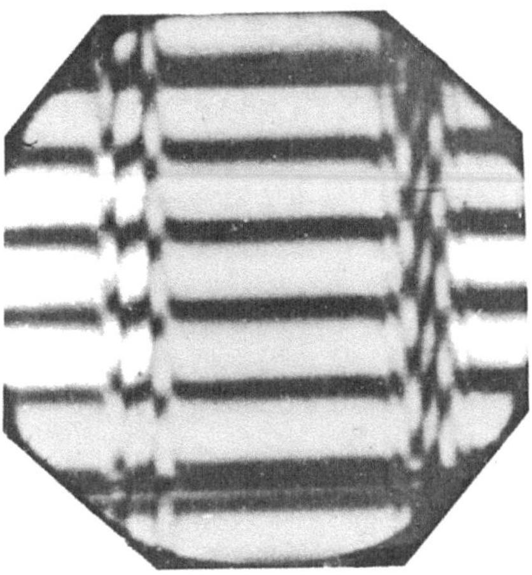

Fig. 47.— Magnetic double Refraction near Components of Sodium Lines.

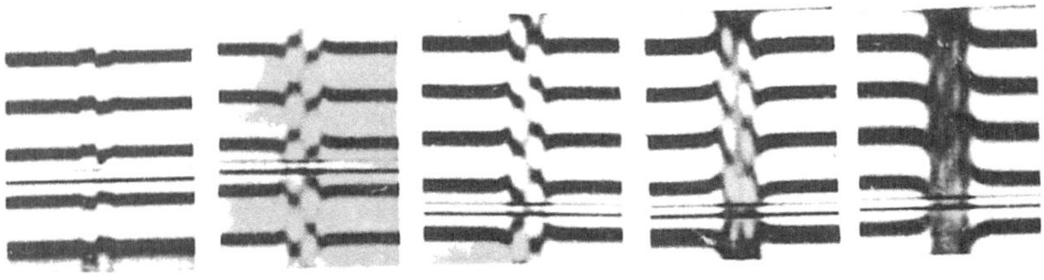

a. 1%. b. 2½%. c. 5%. d. 10%. e. 20%.

Fig. 48.— Lithium Triplet and Magnetic Double Refraction.

When the quantity of sodium increased (always remaining absolutely small, however), the interference fringes moved upward along the components of the doublet; at the same time, the part of the fringe between the components seemed no longer connected to the exterior fringes. The outer parts of the inmost fringe seemed to have been shifted downward.

On further increase of the density of the vapour, the interior part of the fringe slid downward with increasing velocity, and then resembled an arrow with point directed upward, the parts more removed from the central line fading away and disappearing. At last the arrow entirely disappeared owing to the increase of the density of the vapour. The space between the inner components was still faintly and uniformly illuminated. The reversed sodium lines of the arc continued to exist. A further increase of the quantity of sodium obscured the central part more and more. The exterior fringes moved continuously upward while the density was being increased.

In a field of 18,000 units the downward displacement could be followed over a distance of more than double the distance between two fringes, corresponding to a *negative* rotation of above $2 \times 180°$, say 400° or 450°.

In the case of D_2 the phenomena were in the main of the same character. It was, however, characteristic of D_2 that the stage of the almost, or entire, vanishing of the interior fringes was reached sooner than for D_1 with smaller field, the shape of the interior fringe also being different from that observed in the case of D_1. Hence there exists also in this case a difference between D_1 and D_2, a difference already known to exist in the phenomenon of reversal, of the resolution by a magnetic field, and of the rotation outside an absorption band.

When the density of the vapour was maintained as constant as possible, it was ascertained that the negative rotation diminished with *increase* of the intensity of the field. When the strength of the field was raised, *e.g.*, from 18,000 to 25,000 units, this decrease was clearly to be perceived without any measurements being required.

An enlarged reproduction of one of our photographs will enable us to get a clear idea of the phenomenon (Fig. 41, Plate III). For D_1 the band between the doublet lines belongs to the band lying above it. It is exceedingly instructive, and one of the most interesting optical experiments, to observe the deformations of the interference bands with large dispersion while the density of the magnetised vapour is being changed. The great value of the negative rotation and the direction in which it changes with increasing field-intensity may be regarded as a remarkable confirmation of the theory.

That the mode of change of the rotation with increase of the field intensity is really required by theory, can easily be made clear by drawing Fig. 39 for a greater magnetic resolution. The dispersion curves then move apart, and become lower at the place of the original spectrum line.

50. In the foregoing theoretical derivation it was supposed for the sake of simplicity that the absorption indices of the two kinds of polarised light are equal. For a definite frequency n, these indices are in general different. Hence, from a linearly polarised ray an elliptically polarised one is formed on propagation. It can be proved that the axes of the ellipse characteristic for the vibration of light undergo a rotation which is given by equation (1), § 47, p. 77. During the propagation the ellipse gradually assumes the shape of a circle. The angle ψ may always be found in the investigation with a nicol by making the intensity a minimum. In applying the method of the interference bands, we shall meet with different states of polarisation at different distances from the absorption lines; the deformed interference band will be a line of minimum light intensity, not of an intensity zero.

51. In his thesis for the doctorate, Hallo[1] carried out a series of measurements on the magnetic rotation of the plane of polarisation outside the absorption bands for a sodium flame. He made use of the method of the interference bands, which were photographed, and he set himself the task of testing the theory numerically and measuring the parameters which occur in the equations of the theory. For though the considerations of § 47 are adequate to give a qualitative survey of the phenomena, verification of the numerical relations is necessary to decide whether the theory can be accepted in every respect.

The determinations of the parameters are very uncertain, for it is difficult to characterise the circumstances in a flame with precision. About the value of the rotation χ we know that outside the lines of the doublet it must in a first approximation be in inverse ratio to δ^2, if δ represents the distance in wave-lengths from the original line. In accordance with Voigt's theory, Hallo really finds $\chi\delta^2$ approximately constant.

We shall not give here the numerical values of the parameters which Hallo derives from his experiments, as an exposition of the physical significance would lead us too far. We will direct attention, however, to one result. From the results of the experiment on the rotation of the plane of polarisation, and with the introduction of the values of e/m and e, Hallo computed a density of about 10^{-8} for the vapour of his sodium flame. The actual density of the vapour in the flame was, however, much greater, perhaps 10^{-6}. This difference may possibly be accounted for by assuming that not all the sodium atoms participate at the same time in the emission or absorption, but only a fraction of them. A similar conclusion was also drawn by Geest in an investigation on the magnetic double refraction, *cf.* § 56. It seems as if the electro-

[1] Hallo, Dissertation, Amsterdam, 1902. Cf. also *Archiv Néerl.* (2), **10**, 148, 1905.

magnetic system of which the atom consists can only absorb energy in some of its configurations.

52. With a somewhat different arrangement of the experiment of § 48, new phenomena may be obtained, which reveal a difference between D_1 and D_2. If horizontally vibrating light is made to enter the flame, and then to traverse a mica plate with the principal directions at an azimuth of 45°, and after this the quartz wedges and the nicol of § 48, the phenomena observed in the spectroscope are shown in the accompanying photographic reproduction (Fig. 42, Plate III), here published for the first time. It was easy for Voigt[1] to draw the conclusion that with weak absorption and double refraction the phenomenon of D_1 must appear, whereas with greater absorption that of D_2 is exhibited.

53. With an excellent Michelson plane grating of width 13·3 cm., and about 600 lines per mm., Voigt has constructed a spectroscope and succeeded in carrying the observations considerably further than was possible with the earlier appliances. The theoretical resolving power of this spectroscope amounts to 240,000 in the third order, which is still sufficiently bright. Practically, at least 200,000 is reached. I am indebted to Prof. Voigt for the remarkable reproductions in Fig. 43 (Plate III.) In the flame used the sodium was introduced by injection of a dilute Na_2CO_3 solution. The system of quartz wedges was double, one half being related to the other as an object to its image. This is an advantage for some measurements. In the accompanying reproduction (Fig. 43) the border line is to be seen near the edge of the image.

The photographs *a*—*f* refer to solutions with 5 per cent. and 10 per cent. sodium carbonate, and field intensities of 9,100, 14,500, 27,500. The *decrease* of the rotation between the components of the doublet when the field

[1] Voigt, "Neue Beobachtungen über Magneto-optische Erscheinungen in Absorptions-streifen." Göttinger Nachr., 1902, Heft. 5.

intensity increases is strikingly illustrated by the photos (*cf.* §. 48). Further, they show the *increase* of the rotation when the concentration is increased, but the intensity of the field is kept constant. Moreover, the extreme sharpness of the images allows us to follow the behaviour of the line D_2.

54. *Magnetic Double Refraction.*—Another consequence of the magnetic resolution of the spectrum lines and of the selective dispersion predicted by Voigt's theory, but one which is an entirely new magneto-optical effect, is observed at right angles to the lines of force. It con-

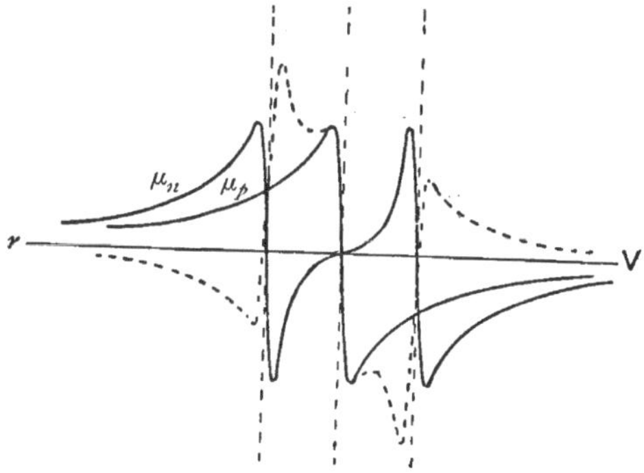

FIG. 44.

sists in a double refraction which is observed when a plane wave traverses a flame in the magnetic field normal to the lines of force. The electric vibrations which in the wave vibrate parallel to the field have a different velocity of propagation from those which are normal to it. A general survey of the expected phenomena may again be obtained by a graphical representation (Fig. 44). For the case of magnetic separation into a triplet, the vibrations parallel to the lines of force are controlled by the central component, whereas the vibrations normal to the field are under the influence of the outer components. For vibrations normal to and parallel with the field, the curves μ_p and μ_n give the indices of refraction as functions of the frequency. The difference of the ordinates $\mu_p - \mu_n$ is indicated by a dotted line. It is seen at a glance that outside the triplet on the two sides the double refraction has opposite signs. Between the components the double refraction is very great and in a direction opposite to that in the adjoining region.

Voigt first succeeded in demonstrating the double refraction predicted by theory for sodium vapour in a direction normal to the magnetic lines of force.

For this purpose he made the light of an electric arc lamp traverse a nicol, the plane of vibration of which formed an angle of 45° with the field, and then placed the flame in it. The light was analysed with a spectroscope. Before the slit the two wedges of quartz of Babinet's compensator were adjusted: behind the slit a second nicol was introduced. As is well-known, Babinet's compensator consists of a combination of two wedges with their planes in the direction of the optical axis. One wedge has its refracting edge parallel, the other at right angles to the optical axis. In monochromatic light between crossed nicols, the quartz system exhibits dark parallel bands, which correspond to simple values of the difference of phase in the wedges.

In Voigt's arrangement, we see accordingly in the field of view of the spectroscope a system of nearly horizontal dark bands at right angles to the vertical spectrum lines. Every difference of phase between vertical and horizontal vibrations will now cause a displacement of the interference bands, and if this difference of phase depends on the wave-length, a deformation of the bands is brought about. In a simple way, the value of the difference of phase for a definite wave-length can be derived from the amount of the vertical displacement. The close analogy which exists between the method discussed above and that of §48 scarcely requires mention. According to theory, the double refraction varies directly with the square of the intensity of the field; hence it does not change its sign if the direction of the field is reversed. This can also be read from the graphical representation in Fig. 44; there is, namely, nothing in the three lines of the triplet and the dispersion curves that changes with reversal of the direction of the field. The deformation of the interference bands must take place, therefore, in opposite directions on either side of the thick absorption line of the sodium vapour, and be independent of the direction of the field.

It is this result of the theory which was confirmed by Voigt's first observations, and proved for the first time the existence of the predicted double refraction. A reproduction of the phenomenon made by the present writer is given in Fig. 45 (Plate IV).

55. Observations of the transversal double refraction, in which the interesting region between the split lines particularly was examined, have been made by Zeeman and Geest.[1] The method used is described in §54, the D-lines of very dilute vapour being investigated with great dispersion. The refracting angles of the compensator wedges were 50′, but for the study of some details compensators were used with angles of 10′ or of 3°.

In strong fields the line D_2 is split up into six components, and the line D_1 into four. The application of the theory to these cases does not, however, present any essential difficulties. Just as for the triplet in Fig. 44, we have to draw the curves representing the indices of refraction μ_p and μ_n as a function of n, and to take the differences of the ordinates.

From Zeeman and Geest's paper Fig. 46 only is reproduced. On the left side is the theoretical curve; on the right side the result of the observation is given.

[1] P. Zeeman and J. Geest, "Double Refraction near the Components of Absorption Lines Magnetically split into several Components," *Proc. Amsterdam Acad.*, 435, Dec., 1904.

The latter is not represented by a photograph, but by a drawing.

In the theoretical curve the steep vertical parts are only sketchily drawn, because they cannot be observed. They are either too fine or they merge into the vertical absorption lines. The image is mainly determined by the thicker horizontal parts of the interference bands.

In the method of observation used, the result of interference bands lying above one another is seen. Thus, the theoretical figure must be considered to be supple-

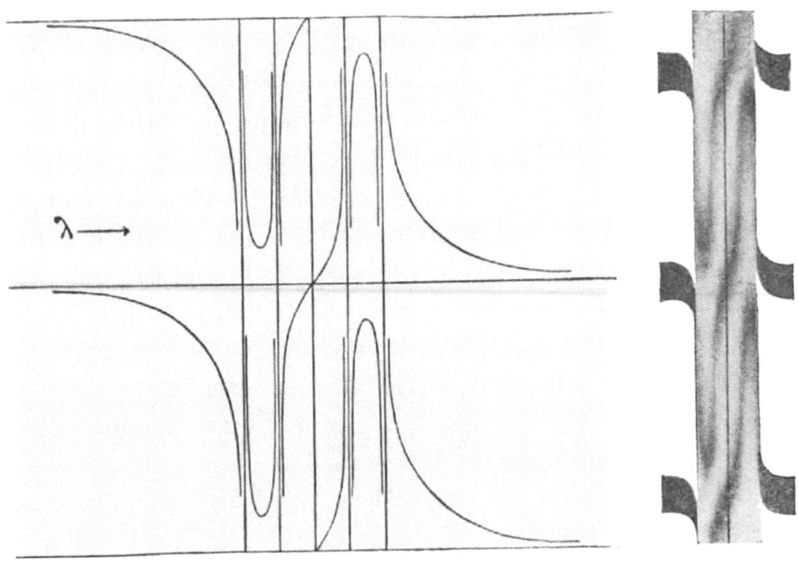

Fig. 46.

mented with parts of theoretical curves which lie in the upper part and in the lower part of the figure. A comparison in detail of the two figures seems unnecessary. For the interpretation of the observations we cannot dispense with the theory.

Characteristic of the double refraction for the quadruplet is the sinuous middle part of the interference band. The double refraction between the outer components greatly exceeds that outside them.

56. In Mr. Geest's thesis for the doctorate[1] the qualitative consideration has been further worked out. Moreover, measurements are given there of the value of the double refraction in sodium vapour at different distances of the absorption band, and the value is computed of some constants which play a part in the theory of the magneto-optical phenomena. These constants are also of some importance for the general optical theory. To one result I will direct particular attention. From the value of the double refraction, Geest calculates a value of the vapour density which is many times smaller than the real density of the vapour in the flame on which he experimented. This result harmonises with that found by Hallo[2] from the magnetic rotation of the plane of polarisation. In § 51 we have already directed attention to a possible explanation of these results by the supposition that all the atoms do not emit light simultaneously. It seems interesting to note here that Lenard,[3] guided by observations on the electric arc, advanced the view that a metallic atom in the arc can exist in a series of different states.

57. The phenomena of double refraction between the components of the split D-lines are only represented by drawings in Zeeman and Geest's investigations. Voigt and Hansen[4] have succeeded in advancing a step further by the aid of their excellent apparatus. For though, as Voigt states expressly, the main features of our drawings are in decided harmony with his photographs, yet the finer details can only be seen and measured in a photograph. One of Voigt's photographs relating to the D-lines is reproduced in Fig. 47 (Plate IV).

No less admirable than this representation is the series

[1] Geest, Thesis for the Doctorate, Amsterdam, 1904. *Cf.* "La double refraction de la vapeur de Sodium," *Archiv Néerl.* (2), **10**, 291, 1905.

[2] Hallo, Thesis for the Doctorate, Amsterdam, 1902, § 51 above.

[3] P. Lenard, *Ann. de Phys.*, **11**, 636, 1903.

[4] Voigt and Hansen, *Phys. Zeitschr.*, **13**, 217, 1912.

of photographs, for which I am also indebted to Voigt, representing the red lithium line (6708 Å.U.). This line is changed into a triplet. The original theory of Voigt, which refers to triplets, applies, therefore, to this case. It is confirmed by Fig. 48, *a–e*[1] (Plate IV), which have been obtained with a flame fed with a dilute solution of lithium, the concentrations of which are given under the figures.

58. There is undoubtedly a peculiar charm in the fact that all the particular properties of sodium vapour in a magnetic field which we have treated in this chapter can be directly reduced to consequences of the joint action of magnetic resolution and anomalous dispersion. The great heuristic value of Voigt's theory for the magneto-optical phenomena in metallic vapours has appeared clearly from what precedes.

There are, however, cases to which Voigt's theory is not applicable. I do not refer here to the magneto-optical effects in crystals, with which we have become acquainted through J. Becquerel's investigations, and those of H. du Bois and Elias. Indeed, for them Voigt and Becquerel developed a theory which closely resembles that sketched in the foregoing sections.

In 1910 Cotton and Mouton[2] discovered the magnetic double refraction of pure *liquids* in observations made in a direction at right angles to the field. This magnetic double refraction, and the analogous property discovered by Kerr for the electric field, cannot be explained by Voigt's theory. According to this theory, the magnetic double refraction is in direct relation with the magnetic rotation of the plane of polarisation; and one phenomenon ought to be calculated from the other. This would also be the case for liquids, if the theory were of general applica-

[1] The figures *c* and *d* have been inverted by mistake.

[2] Cotton and Mouton, *Ann. de Chim. et de Phys.* (8), **19**, 155; **20**, 195, 1910.

tion. But it now appears that the observed double refraction for liquids exceeds the calculated value more than a thousand times.

In the course of their above-mentioned investigation Cotton and Mouton formulated the hypothesis, already proposed by Sir Joseph Larmor for the Kerr electro-optic effect, that in both cases (the electric and the magnetic) the double refraction is due to a directive action of the external field on the molecules of the liquid. The only cause interfering with a parallel arrangement of the molecules is their thermal motion.

In an interesting theoretical investigation Langevin[1] has shown in how simple a manner this hypothesis of the molecular orientation of naturally aeolotropic particles, directed at random in isotropic substances, explains the two phenomena mentioned. By means of this hypothesis of the molecular orientation, Langevin drew up formulæ which are quantitatively corroborated by experience. According to Langevin's theory, the magnetic rotation of the plane of polarisation, and the magnetic double refraction, must be ascribed to different causes. The first phenomenon, closely allied to the magnetic resolution of the spectrum lines and diamagnetism is to be explained by the electromagnetic modification of the paths of the electrons in the atoms through the magnetic field. The second—the magnetic double refraction—to the rearrangement in the orientation of the molecular axes.

The two phenomena are dependent on the temperature in an entirely different way. This had appeared to be the case in the liquids examined by Cotton and Mouton. The magnetic double refraction rapidly decreases with rise of temperature. In agreement with Langevin's formula, it decreases more rapidly than in inverse ratio of the absolute temperature. The magnetic rotation of the plane of polarisation, on the other hand, if calculated on a constant

[1] Langevin, Le Radium, 7, Sept. 1910.

number of molecules, is independent of the temperature, as Langevin's theory requires.

It seems, however, that Langevin goes too far when he desires to explain all the cases of magnetic and electric double refraction by molecular orientation.

The, at first sight, very complex phenomena of double refraction in metallic vapours which exhibit magnetic resolution are explained by Voigt's theory in such a simple and rational way, that Langevin's theory cannot be applicable to them. Nor has anything hitherto appeared of an influence of temperature in the magneto-optical phenomena in metallic vapours.

Thus, according to the circumstances of the case, it is permissible to work either with Voigt's theory or with that of Langevin. It is, however, easily possible that the absorption bands of the liquid are split up by magnetic forces, and that at the same time the molecular orientation changes. If so, we should always have to deal with a combination of two actions, but with only one predominating in most cases.[1]

[1] Cotton gives (in the paper cited, § 58) a somewhat different view of the part played by absorbing molecules on magnetisation.

After the above had been written the following paper by Voigt appeared: Über magnetische und electrische Doppelbrechung, I. & II. *Göttinger Nachr.* 1912. In this Langevin's theory is extended: "Zeeman Effekt und Orientierung der Moleküle wirken m. E. *neben* einander; in manchen Fällen dominiert der eine, in manchen der andere Umstand, derart dass seine alleinige Betrachtung zur Erklärung anscheinend ausreicht. Diesen Standpunkt, den auch Herr Cotton einzunehmen scheint, dürfte die Gesamtheit der Erfahrungen gegenwärtig an die Hand geben, und *ihn* zu bezeichnen ist der eine Zweck meiner Mitteilung."

CHAPTER VI

INFLUENCE OF THE GRATING AND THE SLIT ON THE INTENSITIES OF THE COMPONENTS; PURITY OF THE CIRCULAR POLARISATION

59. In our treatment of the laws of magnetic resolution the geometric relations between the components were always put in the foreground. It is, however, of great importance that the ratio of the intensities of the split components, and the purity of polarisation of the emitted light, should also be taken into account. Knowledge of the ratios of the intensities is indispensable in a theory of the complex resolutions.[1]

As yet, this problem has been only scantily treated experimentally. In many cases the grating itself, as well the slit, can exert a strong influence on the ratio of the intensities of the components. These points were examined in our own laboratory,[2] and refer to disturbances which must be elucidated before the more essential question can be taken in hand.

In the first part of this chapter we shall discuss *the influence of the grating and the slit on the intensities of the components*; and in the second part, *the degree of completeness of the circular polarisation.* The result of this latter part leads to a limitation of the possible paths of the

[1] Voigt, *Ann. d. Phys.*, **24**, 193, 1907. Göttinger Nachr., 1911.

[2] Zeeman, "The Intensities of the Components of Spectrum Lines Divided by Magnetism," *Proc.* Amsterdam Acad., Oct., 1907.

electrons. Both parts serve to define our former results more closely.

I. *Influence of the grating and the slit on the intensities of the components.*

60. If a spectrum line is resolved into a triplet by the application of a magnetic field, the two outer components and the middle line will generally differ in intensity. According to the elementary theory of Lorentz, there exists a simple relation between these intensities. Let I_1 and I_3 be the intensities of the outer components, and I_2 that of the middle line, then we may expect that

$$I_1=I_3=\tfrac{1}{2}I_2 \quad . \quad . \quad . \quad . \quad . \quad . \quad . \quad . \quad . \quad . \quad (1)$$

For if we first consider a triplet the components of which lie apart, and if then the intensity of the field be supposed to approach zero, the three components will finally coincide, and the line must be unpolarised. For the latter, it is required that the relation (1) is satisfied. This is, however, often not the case. Lines for which in the field relation (1) does not hold, will exhibit partial polarisation when they are examined with a spectroscope of feeble resolving power. This was actually found by Egoroff and Georgiewsky.

61. There are, however, cases which are only apparently in contradiction with relation (1). It has been known for a long time that a grating reflects vertical and horizontal vibrations differently. That, however, this influence could be so strong that the ratio of the intensities for a triplet was quite reversed, was brought to light for the first time by some experiments made by the present writer. With a large Rowland grating, the transversal effect for the yellow mercury lines was observed in the spectrum of the first order. The incident rays made an angle of about 19° with the normal to the grating; and in this latter direction photographs were taken. The

magnetic field was, as usual, horizontal, and the slit vertical. In Fig. 49*a* (Plate V), a reproduction is given of the triplet of the mercury line 5769·4. The distribution of intensity is in absolute contradiction to equation (1).

Light of an ordinary sodium flame, on which no magnetic forces act, was then thrown on the grating at the same angle of incidence as before. Observing in the direction of the normal, I now saw with a calcspar prism that the light was strongly polarised. The vertical vibrations strongly preponderated. The direction between the vibrations and the lines of the grating has accordingly a strong influence on the intensity of the reflected light. If, therefore, the vibrations in the components of the triplet are rotated by the introduction of a quartz plate before the slit, a marked change must take place in the triplet. In the photograph reproduced in Fig. 49*b*, a quartz plate which rotates the plane of polarisation through an angle of about 90°, was placed before the slit. The outer components are scarcely visible. Fig. 49*c* was obtained by means of a quartz plate, 2·15 mm. thick, for which the rotation of the plane of polarisation amounted to $22{\cdot}72 \times 2{\cdot}15 = 48°90$. The distribution of light seems no longer in contradiction to equation (1). A quartz plate which rotates the plane of polarisation through an angle of exactly 45°, and which is placed before the slit, causes the circumstances, so far as the direction of vibration is concerned, to be the same for each of the components.

For the knowledge of the actual ratio of the intensities of the components of a split spectrum line, the slit of the spectrum apparatus must make an angle of 45° with the incident vibrations. If a beam which is to be studied is rectilinearly polarised, it may be advantageous to change the direction of the vibrations by means of a quartz plate, so that they are reflected and diffracted as strongly as possible.[1]

[1] Cotton, "Le phénomène de Zeeman," 21.

That in a number of cases equation (1) is not satisfied can be directly observed in some spectra which are rich in lines (*e.g.*, iron, Fig. 20, Plate I). Triplets lying close together exhibit reversed ratios of intensity. A polarising action of a grating or glass prism cannot account for this; there is a real deviation from equation (1).

62. Besides the polarising effect of the grating, there is a second cause tending to make the ratio of the intensities of components of different direction of vibration in the image different from that corresponding to the constitution of the emitted light. I mean the polarisation impressed upon light which traverses fine slits. Since Fizeau, this effect has been well-known, but the errors which may ensue from it in the investigation of spectrum lines magnetically resolved were first pointed out by the present author.[1]

With a spectroscope securing great illumination, the slit of which is within the observer's control, and with the yellow and the green mercury lines, it is easy to observe the following phenomena when the transverse effect is examined. If the slit be narrowed, the intensity of all the components of the split lines decreases, but that of the central component or group more than that of the outer components. It is even possible to make the central lines of the triplet, or the central group of the green line, disappear entirely. This can only be explained by considering that vibrations perpendicular to the slit scarcely traverse the narrow slit.

A few control experiments can be made easily. If a quartz plate rotating the plane of polarisation through 90°, be introduced before the slit of the spectroscope, it is only the outer components that can be caused to disappear. A second observation can be made with the slit only,

[1] P. Zeeman, "On the Polarisation Impressed upon Light by Traversing the Slit of a Spectroscope and some Errors Resulting therefrom," *Proc. Amsterdam Acad.*, October, 1912.

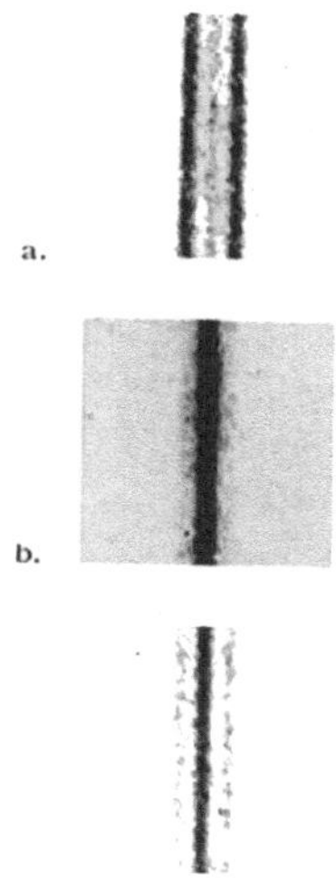

Fig. 49.

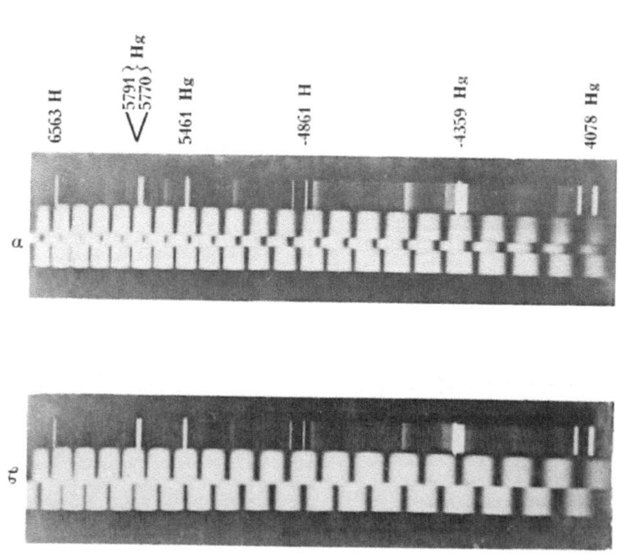

Fig. 50.

which is examined through a calcspar rhomb. If the slit is sufficiently narrow, only the image formed by the vertical vibrations remains visible. The polarisation caused by a too narrow slit gives rise to errors in determinations of intensity of components, and also to apparent shifts and dissymmetrical resolutions of spectrum lines with originally dissymmetrical distribution of light. A quartz plate introduced before the slit of the spectroscope, giving a rotation of the plane of polarisation of 45°, eliminates the source of errors just as in § 61.

II. *The Degree of Completeness of the Circular Polarisation of Magnetically Divided Lines.*[1]

63. The lines of the doublet which is observed in the direction of the magnetic lines of force are circularly polarised (see Chapter II); one component left-handed, the other right-handed. This circular polarisation is, as my first experiments already showed, almost perfect. Lorentz's elementary theory led us to expect perfect circular polarisation. It would, however, not be impossible that a theory which entered more deeply into the phenomena, and accounted also for the complicated resolutions, should arrive at another result.

In view of the great difficulties which a theory of the complicated forms of the magnetic effect has to face, it is interesting to note that certain conclusions concerning the polarisation of the components can be drawn from general principles independently of any particular theory. In 1898, Lorentz[2] concluded from general principles concerning the action of the electro-magnetic forces that light radiated

[1] P. Zeeman, "The Degree of Completeness of the Circular Polarisation of Magnetically Divided Lines," *Proc. Acad. Amsterdam*, October, 1909. This paper may be consulted for other details than can be given in the present chapter.

[2] Lorentz, *Proc. Acad. Amsterdam*, June 25th, 1898, 113. "The Theory of Electrons," Teubner, Leipsig, 130, 1909.

along the lines of force, can never show a trace of linear or elliptic polarisation; it must either be unpolarised, or have a circular polarisation, partial or perfect.

In 1900, Larmor[1] concluded that perfect circular polarisation of the components of the doublet would prove that the corresponding permanent types of vibration in the molecules are exactly circular.

In order to explain the more complicated magnetic effects, Voigt[2] specialised the general theory of vibrating electric systems by supposing particular magnetic connexions between electrons, and by introducing the hypothesis that the luminous particles take a different orientation under the action of the field (see, further, Chapter *X*). For a particular direction—the "axis" of each particle—becomes, under the action of the field, parallel to the lines of force. A rotation of the particles around this axis is undoubtedly possible. This rotation has no influence upon the frequency; the orbits of the electrons, however, are rotated, as are also the orbits of the "equivalent" electrons recently introduced by Lorentz[3] in order to simplify the theory of systems containing a number of electrons. A single vibrating "equivalent" electron with its state of equilibrium somewhere in the radiating atom can take the place of the whole system of which the atom consists. It may be concluded from the linear polarisation of the light radiated perpendicular to the lines of force that the paths of the electrons are straight lines parallel to the lines of force, or ellipses in planes perpendicular to the field. Partial circular polarisation would prove the existence of ellipses with all possible fortuitous orientations in planes normal to the field. The completeness of the circular polarisation parallel to the magnetic force would prove that the ellipses are circles.

1 Larmor, "Aether and Matter," 345.
2 Voigt, "Magneto-und Elektro-optik," Teubner, Leipsig, 98 ff., 1908.
3 Lorentz, "Encycl. Math. Wiss." **3**, Heft. 2, 204, 217.

In general, it would be expected that the components of a magnetically subdivided line would emit partially polarised light parallel to the lines of force.

64. I have made a quantitative investigation of the percentage of the total light, completely circularly polarised, emitted by the components. As will appear, *all* the light radiated parallel to the lines of force is circularly polarised. This was also the conclusion formerly drawn from my observations with Rowland's grating. Considering the slight intensity of the diffraction spectra, however, it was still easily possible that a not insignificant fraction of the intensity of the spectrum line was unpolarised, but remained below the limiting value necessary for perception. With Michelson's echelon spectroscope, which as to brightness surpasses all other spectroscopes of high revolving power, the circumstances are much more favourable.

Sodium, mercury, and thallium present spectra of great intensity. These few elements, moreover, have the advantage of exhibiting several different types of magnetic separations.

We shall now give some particulars of the verification of circular analysers, which, of course, are of importance for the investigation under consideration.

65. *Verification of the circular analysers.*—In order to investigate the circularly polarised light of the components, it is simplest to use either quarter-wave plates or Fresnel's rhomb.

The extent of the double refraction caused by a mica plate may be estimated by the colour it exhibits when it is placed between crossed nicols in parallel light. A closer estimation than that given simply by the colour is obtained when the light leaving the second nicol is analysed by the aid of a spectroscope. The most accurate result is reached if a moderately thick quartz plate cut parallel to the axis is combined with the mica plate. The simple theory of the dark bands now visible in the spectrum is well known, and does not require to be stated here. Since the time of Fizeau and Foucault, many physicists have used these bands for the measurement of phase differences.[1]

These bands are most distinct when the principal sections of the nicols and the mica are inclined at an angle of 45°. If the mica and the quartz plate are superposed in such a manner that the principal sections correspond, a displacement of the bands in a certain sense, *e.g.*, towards the red, is observed. If the mica be then rotated through 90° the displacement is towards the violet.

The ratio of the displacement of the band from its original position, and of the distance between two succeeding bands, gives the difference of phase produced by the mica for the region of the spectrum under consideration. For greater accuracy, I have sometimes used mica plates divided into three fields with horizontal lines of separation. The principal sections of quickest vibration in the outer fields incline 45° to the right; in the central field 45° to the left. An image of the horizontal lines of separation is projected by means of an *achromatic* lens upon the slit of a spectroscope.

Fig. 50*a* gives such a photograph taken with a three-fold mica plate. How largely the phase-difference depends upon the wave-length is clearly shown. In the red region of the spectrum the plate is almost exactly a quarter-wave plate.

As is well known, Fresnel's rhomb gives circularly polarised light after two total reflections.

In order to obtain two fields, a double plate of quartz may be used. I used a quarter plate, 1·7 mm. thick, cut from the crystal parallel to the axis. This plate is cut into two halves by a line making angles of 45° with the

[1] A. Cornu, *Compt. rend.*, **125**, 555, 1897 ; "Eclairage Electrique," **13**, 246, 1897, A. Cotton. *Ann. de Chim. et de Phys.* (7), **8**, 1896

principal section ; one-half is then rotated round an axis perpendicular to the line of separation, through 180°.

Fig. 50*b* is a reproduction of a photograph taken with a rhomb. The difference as compared with using a quarter-wave mica plate is very apparent.

The error in the phase-difference of 90° which exists between the two linear components into which circularly polarised light can be resolved amounts to less than 5·°4. This error is, further, without influence on the determinations. If the rhomb does give a phase difference which deviates by a small angle δ from 90°, very feebly elliptically polarised light emerges in the case of incident circularly polarised light. The ratio of the axes of the characteristic ellipse is, as easily appears, $\frac{1}{2}\sin\delta$, and for $\delta = 6°$ this becomes 0·0522.

The intensity of the light leaving a nicol with its plane of vibration perpendicular to the major axis of the ellipse is $(0{\cdot}0522)^2 = 0{\cdot}0027$. As the minimum intensity which in the circumstances of our observations may be recognised will appear to be of the order 0·01, even an error twice or three times that of Fresnel's rhomb cannot vitiate the results.

66. After this digression upon circular polarisers a brief account may be given as to the arrangement of the experiments on complete polarisation.

The light emerging parallel to the lines of force of a du Bois' electromagnet (with one perforated polar piece) was made parallel by means of an achromatic lens, traversed Fresnel's rhomb, then a second achromatic lens, which forms a sharp image of the source of light on the slit of an auxiliary spectroscope. Between the second lens and the slit of the auxiliary spectroscope, a nicol was introduced, which could be made to rotate about its axis. The degree of the rotation could be read on a graduated circle. The front plane of the rhomb was placed accurately perpendicular to the incident beam of light.

It then appeared that there was no trace of ordinary light in the doublets—sometimes divided further. One of the components could be quenched entirely by means of the nicol. The value of this result depends on the sensitiveness of the method.

If we start from the position of the nicol in which no light is perceived, the angle a may be determined at which light is first seen again.

If a be the angle the nicol is thus rotated from the zero position, $I \sin^2 a$ represents the brightness of the emergent light, I being the intensity of the linear vibration.

We can be sure that the quantity of ordinary light emitted by one of the components must be below $I \sin^2 a$. Vacuum tubes were used as the source of light. For further particulars we refer to the original paper.

67. The observations consist in determining the value of a for different spectrum lines. The results may be arranged according to the different types of subdivisions, observed normally to the field.

Triplet, mercury, $\lambda = 5791$, doublet parallel to the field, $a = 7°$ $\sin^2 a = 0{\cdot}0144$. The observation is somewhat hindered by satellites of the principal lines.

Triplet, $\lambda = 5771$, doublet in direction of field.

In extremely strong fields, every component is resolved into three lines. This strength of field was not reached.

$$a = 5° \quad \sin^2 a = 0{\cdot}0076.$$

Quartet, sodium, $\lambda = 5896$, doublet in direction of field.

$$a = 5° \quad \sin^2 a = 0{\cdot}0076.$$

Sextet, sodium, $\lambda = 5890$, quartet in direction of field.

$$a = 6° \quad \sin^2 a = 0{\cdot}0108.$$

Nonet, mercury, $\lambda = 5461$, strong green mercury line, sextet in direction of field.

$$a = 5° \quad \sin^2 a = 0{\cdot}0076.$$

Sextet, thallium, $\lambda = 5351$, quartet in direction of field.

$$a = 8° \quad \sin^2 a = 0{\cdot}0196.$$

In the last observation the tube was not in a satisfactory condition.

The experiments described in the latter half of this chapter have proved that the spectrum lines of a gas in a magnetic field emit only circularly polarised light along the lines of force, and that, at least for several line spectra, the amount of ordinary light which may be emitted simultaneously with the circularly polarised light is less than 1 per cent. of the total intensity of the spectrum line.

We must conclude that the orbits of the equivalent electrons, in planes normal to the magnetic force, are, with close approximation, circles. Elliptic orbits, fortuitously distributed in planes normal to the field, need not be conjectured for the representation of the phenomena.

Somewhat closely connected with the present subject are some investigations of Jean Becquerel[1] and A. Dufour.[2] Dufour obtained various new results concerning the banded emission spectra of the alkaline-earth fluorides and chlorides radiating in a magnetic field, and in some cases observed incomplete circular polarisation. According to J. Becquerel, several absorption bands of xenotime and tysonite also exhibited incomplete circular polarisation in a longitudinal magnetic field. He showed,[3] however, that in the case of these crystals there is no real incomplete polarisation, but that, under the action of the magnetic field, besides the principal components, others of slightly different wave-length come into existence, and exhibit a polarisation opposite to that of the principal lines.

[1] Jean Becquerel, *Compt. rend.*, **145**, 413, 1907.

[2] A. Dufour, *Compt. rend.*, **146**, 118, 229, 1908. *Journal de Physique*, April, 1909.

[3] Jean Becquerel, *Amsterdam Proc.*, 146, June, 1909. "Contribution à la connaissance du phénomène de Zeeman dans les cristaux." See also supplement No. 20. Leyden communications.

CHAPTER VII

DISSYMMETRIES AND SHIFTS

68. MANY of the stronger lines, for instance those of the iron spectrum, seem to furnish an ideal example of the elementary theory. It is, however, none the less remarkable that the simple distribution of intensities and distances is in agreement with the elementary theory.

The question may be raised whether external magnetic forces do not favour the circular vibrations of the electrons more than those along the lines of force. The molecular circular currents, to which para- and dia-magnetism are attributed, are directed or produced by an external field ; and it can very well be characteristic of the structure of the radiating particles that the revolution of the electrons is promoted. If, however, this is accepted, we must also come to the conclusion, as Lorentz has observed, that the revolution of the electrons in one direction must take place to a greater degree than in the opposite direction. A difference of intensity must thus be expected both between the outer components of the triplet and between the two components of the doublet.

In a paper of June, 1898,[1] I tried to answer the question whether dissymmetry of the intensities is found in strong magnetic fields.

In a paper on the partial polarisation of light radiated

[1] "On an Asymmetry in the Change of the Spectrum Lines of Iron, Radiating in a Magnetic Field." *Proc. Acad. Amsterdam*, June, 1898.

by a source of light in a magnetic field Lorentz derived, from calculations which are in direct relation with the elementary theory, that the differences of intensity between the components can be very trifling only.[1] At least, this is the case if we accept the assumptions of the simple theory which have been already set forth. It is, however, possible that other assumptions would lead to a greater dissymmetry.

This rendered desirable an experimental inquiry into the existence of dissymmetry. It may also be assumed from Lorentz's formulæ for the amplitudes that when a directing influence exists as above mentioned, the component on the side of the greater wave-lengths will be the stronger one, both in the triplet and in the doublet, a result which is independent of the sign of the charge of the electrons.

The first results obtained with the spectrum of an iron-spark in the magnetic field were very promising. Several iron-lines with wave-lengths between 3000 and 4000 Å.U. gave triplets dissymmetric with regard to the intensity in strong fields. The investigation was made by means of a Rowland grating in the spectrum of the second or third orders. I could, however, prove that several of the dissymmetries were due to disturbances. As a rule they were caused by the superposition of a feeble component of a neighbouring line on one of the components of the triplet under consideration. My conclusion was that in the iron spectrum a directing influence of the external field on the radiating particles does not reveal itself—at least not in the ratios of the intensities of the triplet lines of the iron spectrum.

69. Of an entirely different nature is the dissymmetry now to be discussed, which was derived by Voigt[2] from

[1] Lorentz, "Verslagen," *Akad. Amsterdam*, 213, October, 1897.

[2] W. Voigt, "Über eine Dissymmetrie der Zeemanschen normalen Triplets," *Ann. d. Phys.*, **1**, 376, 1900.

his theory of the inverse effect treated in Chapter V. According to Voigt, normal inverse triplets in *weak* magnetic fields exhibit a dissymmetry of the following nature: the outer component on the side of the *red* has the *greater intensity*, and the component on the side of the *violet* will be at the greater distance from the original line. In weak fields these dissymmetrics will preponderate upon observation perpendicular to the field, disappearing, however, in strong fields. Fig. 51 represents the theoretical change of the triplet with increasing intensity of the field. The intensity of the absorption is indicated by the thickness of the line; *R* denotes the direction towards the side of the greater wave-lengths.

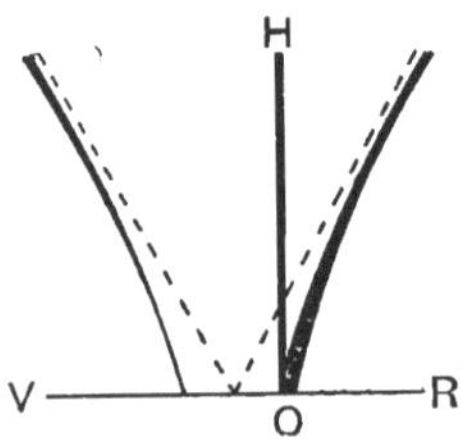

FIG. 51.

It is very remarkable that the two absorption lines do not originate in the resolution of the original line, but by the appearance of a new line at the side of the original one. With increasing field, the new component gains in intensity while the other diminishes, both receding symmetrically with respect to their original middle. With great intensities of the field, the dissymmetry of the intensities disappears, and the ratio of the distances of the components to the central line approaches unity.

According to theory, the doublet which may be observed in the direction of the magnetic force must be due to a symmetrical resolution of the absorption line. This leads us to the somewhat curious result that the doublet of the linear vibrations in a direction perpendicular to the field emits other wave-lengths than the longitudinal doublet.

For a single vibrating electron, this result will be absurd. We have to deal here, however, with a very great number of electrons, which are subjected to an interaction on each other. This interaction is entirely different in the two principal directions, in consequence of which the remarkable result of the theory appears somewhat less strange.

70. At the end of 1899, at Voigt's suggestion, I undertook[1] some observations on the dissymmetry in weak fields to test the results of his theory. The observations were made with emission lines, to which the theoretical results obtained for the absorption may, no doubt, be applied. The Rowland grating used had a radius of about 3 m. and was ruled with 14,438 lines per inch, having a total number of about 50,000 lines. In the spectra of the second and third orders, and for wave-lengths between 3400 and 3900, photographs were taken of an iron spark. For some ten lines, which split up into triplets, a dissymmetry of the distances or of the intensities, and sometimes of both, could be observed; but the phenomenon was exceedingly faint. Some quadruplets were also examined. Two showed a slight displacement towards the violet of the middle of the outer components with regard to the middle of the inner. Voigt's developments, indeed, only refer to triplets, but nevertheless, it does not seem unwarrantable to consider these observations on quadruplets as indications in favour of the theory.

For a single iron triplet, the component on the violet side was found at a *smaller* distance from the original line than that on the side of the red. In a recent excellent investigation by King, these results were corroborated and extended.[2] Fourteen lines in the iron and titanium spectra show dissymmetry, either in the spacing of the components or in the intensities of the violet and red components.

71. From the foregoing observations it seemed justifiable to accept the real existence of dissymmetries. The

[1] "Some Observations concerning an Asymmetrical Change of the Spectrum Lines of Iron, Radiating in a Magnetic Field," *Proc. Acad. Amsterdam*, December 30th, 1899.

[2] A. S. King, "The Influence of a Magnetic Field upon the Spark Spectra of Iron and Titanium" (Carnegie Institution Papers of the Mount Wilson Solar Observatory, **2**, Part I), 1912.

extreme minuteness of the dissymmetry, however, made it desirable to establish its reality in another way.

A welcome confirmation of the results obtained was furnished by an investigation by Reese,[1] who found deviations in the same sense as I had obtained, in experiments not expressly undertaken with the view of observing dissymmetries. In one respect I thought I could improve upon the method followed in my first experiments on dissymmetry. The negatives were taken with a concave grating which was mounted according to Rowland's method. The spark passed in a homogeneous magnetic field, photos being taken with different field intensities. There are then, however, always but very few triplets that have just been split up, a circumstance in which the dissymmetry is first to be expected.

It seems, therefore, more promising to use an extensive source of light, which is exposed in different parts to unequal magnetic forces. By projecting an image of the source of light on the slit of the spectroscope, the spectra of its different parts can then be investigated simultaneously, if only care be taken that the spectroscope be stigmatic.

Attention may here be directed to the fact that it is thus rendered possible by the magnetic resolution of spectrum lines to measure *the magnetic intensity of the field simultaneously in all the points of a straight line.* For it has appeared from the experiments that the resolution of the spectrum lines is directly proportional to the intensity of the field in which the source of light is placed. As the resolution can be measured very accurately, it is possible to derive inversely from this the intensity of the field in absolute measure, when once the specific resolution is known. In many cases[2] it can be given with a probable error which comes to a good deal less than

[1] Reese, *Astrophysical Journal*, **12**, 134, 1900.

[2] Compare the measurements by Weiss and Cotton and by Gmelin, § 40 above.

1 in 100. A *comparison* of field intensities by means of resolution is of course possible without the knowledge of the specific resolution.

It is now of much importance for the study of dissymmetry to be able to study the phenomenon in fields of different intensity in a single diagram. The method that I have now set forth may be called *the method of the non-uniform field.*

We are then certain to work with different intensities, but for the rest in constant circumstances. I first described the method in a paper on magnetic resolution of spectrum lines and magnetic force.[1]

The concave 6-inch grating used had a radius of curvature of 6·5 m. and 10,000 lines per inch, and was stigmatically mounted. The observations were made in the spectrum of the first order. A vacuum tube with mercury was the source of light, which was placed in the non-uniform field. Fig. 52 (Plate VI) gives a ninefold enlarged reproduction of the yellow Hg lines. So far as can be seen, the left-hand line $\lambda = 5770$ is split perfectly symmetrically, the right-hand line $\lambda = 5791$, on the contrary, is clearly dissymmetric. The sense of the dissymmetry of the distances would be in agreement with theory, which seems also to be the case for the intensities, though not to such a marked degree.

The resolution of the line 5770 appeared to be so very nearly symmetrical that it could be used as a measure of the magnetic force.

Measurements were made for intensities of the field between about 14800 and 29220. In order to give an idea of the amount of the dissymmetry, I may state that to a resolution of the line 5770 towards red and violet of 0·532 Å.U. (to which an intensity of the field corresponds of about 29200), correspond the resolutions 0·475 Å.U.

[1] Zeeman, "Magnetic Resolution of Spectrum Lines and Magnetic Force," *Proc. Acad. Amsterdam*, April, 1906; November, 1907.

towards red and 0·523 Å.U. towards violet for the line 5791. Hence the absolute amount of the dissymmetry is equal to 0·048 Å.U.

The absolute amount of the dissymmetry, *i.e.*, the difference of the resolutions towards red and towards violet, appeared not be constant, but to depend on the intensity of the field; it was found approximately proportional to the intensity of the field (see however § 78).

A comparison of the experiments with theory will be discussed later.

72. Though it seems very improbable that the results of the dissymmetric resolution are to be attributed to errors in the Rowland grating, some doubt in this direction is still possible.

The method of the semi-silvered plates of Fabry and Perot seemed very suitable for control experiments; moreover it permits an extension of the investigation to weak fields.

The simplest and least expensive form in which the method of the parallel semi-silvered plates can be applied, and in which it was used for the measurements of wave-lengths by Fabry and Perot, Lord Rayleigh, Eversheim, and others, is the étalon. The distance of the silvered plates is then very nearly constant. The experimenter himself can secure the perfect parallelism of the plates as the theory of the apparatus requires.

73. The theory of the comparison of wave-lengths by means of this apparatus is simple, and has been given by Fabry and Perot. We will apply it to the magnetic resolution of spectrum lines, and especially to the simplest case—the division into a triplet.

Let λ_0 be the wave-length of the original line (afterwards, therefore, the middle line of the triplet). To this corresponds a system of rings; let P_0 be the ordinal number of the first from the centre. The ordinal number at the centre p_0 is then the integer number P_0, aug-mented with a fraction ϵ_0, hence $p_0 = P_0 + \epsilon_0$. Ordinarily $_0 < \epsilon_0 < 1$.

The diameter of a ring increases with ϵ. Let e be the thickness of the plate of air; the order of interference at the centre is $p_0 = \frac{2e}{\lambda_0}$. At an angle i with the normal to the plate the order of interference is $p_0 \cos i$.

Let x_0 be the angular diameter of the ring P_0; then we have, observing in the focal plane of a lens, $p_0 \cos\frac{x}{2} = P_0$. Developing the cosine

$$p_0 = P_0\left(1 + \frac{x_0^2}{8}\right)$$

or

$$\epsilon_0 = P_0 \frac{x_0^2}{8} \quad . \quad . \quad . \quad . \quad . \quad . \quad . \quad . \quad . \quad . \quad (1)$$

Let λ_r be the wave-length of the outer component of the triplet towards the red, then, if P_r, ϵ_r and x_r have a significance corresponding to that of P_0, ϵ_0 and x_0,

$$\epsilon_r = \mathrm{P}_r \frac{x_r^2}{8}.$$

We have, however, $\lambda_0(P_0 + \epsilon_0) = \lambda_r(P_r + \epsilon_r)$, whence

$$\lambda_r = \lambda_0 \frac{P_0}{P_r}\left(1 + \frac{x_0^2}{8} - \frac{x_r^2}{8}\right) \quad . \quad . \quad . \quad . \quad . \quad (2)$$

In like manner, λ_v, P_v, x_v determining the component of the triplet towards the violet, we have

$$\lambda_v = \lambda_0 \frac{P_0}{P_v}\left(1 + \frac{x_0^2}{8} - \frac{x_v^2}{8}\right). \quad . \quad . \quad . \quad . \quad (3)$$

In the case of radiation in a magnetic field, this expression may often still be simplified. In many cases we may choose

$$P_0 = P_v = P_r \quad . \quad . \quad . \quad . \quad . \quad . \quad . \quad . \quad . \quad (4)$$

Looking at the system of rings corresponding to λ_0, while the magnetic force is slowly but gradually increased, we see at the same time rings which proceed from the system λ_0 and are moving outwards and others which are moving inwards. The rings corresponding to λ_r are contracting, those corresponding to λ_v are expanding.

It depends upon the value of p of the étalon, and upon the intensity of the magnetic field, how great will be the expansion and contraction of the rings, in comparison with the distance of the rings λ_0. The value of p and the maximum magnetic force will determine whether in the centre new rings will respectively appear or disappear.

We need not select for measurement the smallest rings, but if the rings λ_r and λ_v, *which originate from the same ring* λ_0, are suitable to be measured, ϵ can become larger than unity.

When we select the rings thus specified the equality (4) applies, and then we may determine λ_r and λ_v from the angular diameters of the rings and the value of λ_0, regarded as known; the result is then independent of the accurate value of the thickness of the plate of air.

Of course the position of the new rings between the rings λ_0 will, with a given value of the magnetic force, be determined by the thickness of the plate of air; and what might be called the "sensibility" of the system of rings to magnetic forces will increase with the thickness of the plate of air. A limit to this sensibility is attained (often too quickly) by the effective width of the spectrum line under consideration.

In some cases, it will be desirable to select for measurement rings different from the three specified. There are no difficulties about the significance of P; it always means the ordinal number of the measured ring. However, if P_0 differs from P_r or P_v, their values must be known for the calculation according to (2) and (3).

74. Fig. 53 (Plate VI) gives an idea of the aspect of the magnetic resolution of the spectrum lines, as observed by means of the method of Fabry and Perot. It is approximately a sixfold enlargement of a negative taken with an

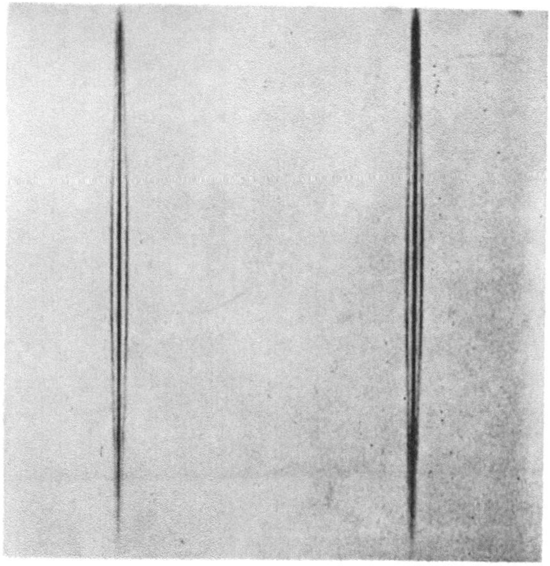

v r

5770 5791

Fig. 52.—Yellow Mercury Lines in Non-uniform Field. 5791 Exhibits Dissymmetry.

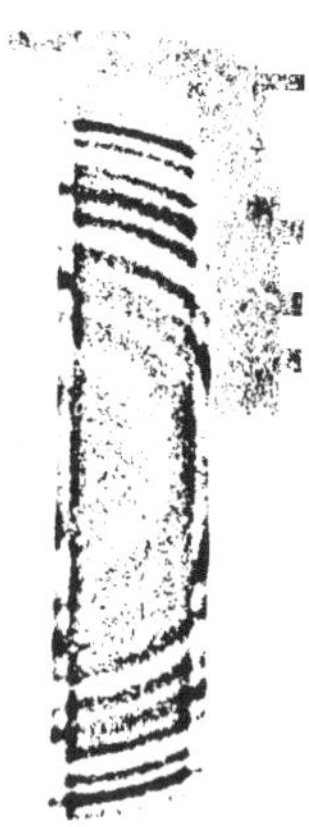

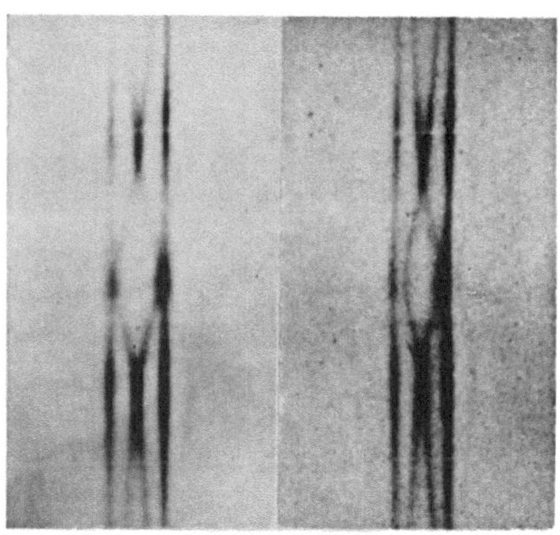

5770 5791

Fig. 54.—Shift of Middle Line of Triplet 5791.
1 mm.=0,12 Å.U.

Fig. 53.

étalon having an interval of nearly 5 mm. between the plates. The mounting and the plates of the étalon are by Jobin. The source of light in the magnetic field was a small vacuum tube charged with mercury. The order of interference at the centre for the mercury line 5791 is about 17265·7 at 16°.

The system of rings was formed in the focal plane of a small achromatic lens of 11 cm. focus. The focal plane coincides exactly with the slit of a small spectroscope. When the slit is opened wide, each spectrum line is seen as a rectangle with bright rings or parts of rings, as the case may be. The figure refers to the yellow mercury lines, the two rectangles corresponding to the separate lines being superposed. For measurements it is desirable to reduce the width of the slit. The intensity of the magnetic field of the figure was about 5000 gauss. It is extremely interesting that so small an instrument as the étalon used, even with the low field mentioned, shows the essential features of the phenomenon.

75. For measurements, to which I shall have occasion to refer later, I have used, not only the method of diameters summarised above, but also the *method of coincidence.* The magnetic force is regulated in such a manner as to bring to coincidence the rings for λ_r and λ_v, or those for λ_r and λ_v with λ_0. By variation of the current in the electromagnet, the coincidence can be attained with the desired degree of accuracy.

In applying the method of coincidences[1] the simplest procedure is to consider the ring formed by superposition of two other rings, once as a ring λ_v derived from a smaller ring λ_0, and again as a ring λ_r derived from a larger one λ_0.

By measuring three rings, viz., that due to the coin-

[1] For further details see: "Observations of the Magnetic Resolution of Spectrum Lines by means of the Methods of Fabry and Perot," *Proc. Acad.* Amsterdam, 440, 1907.

cidence of the rings λ_r and λ_v (diameter $x_c = x_r = x_v$), then the larger ring with diameter x_0, and finally the smaller ring with diameter x'_0, the result may be found by the simple formulæ:

$$\lambda_r = \lambda_0\left(1 + \frac{x_0^2}{8} - \frac{x_c^2}{8}\right),$$

and

$$\lambda_v = \lambda_0\left(1 + \frac{x'^2_0}{8} - \frac{x_c^2}{8}\right).$$

76. Using the 5 mm. étalon, by means of the method of coincidences I made some measurements of the magnetic resolution of the yellow mercury lines.

In the paper cited I obtained, as a result of six measurements concerning the first coincidence, that to a field intensity of 9130 gauss corresponds a separation of the line 5791 towards the red of 0·160 Å.U. and towards the violet of 0·177 Å.U. The numerical value of the separation would amount, therefore, to 0·017 Å.U. for the field under consideration. A discussion of the systematic errors of observation to be feared shows that it is yet possible that the value of the dissymmetry is 0·015 Å.U. or 0·019 Å.U., but that the values 0·011 Å.U. or 0·023 Å.U. are very improbable.

The primary object of our experiments with the method of silvered plates was not so much the exact determination of a numerical value, as the independent verification of results obtained by Rowland's method.

77. According to theory, the doublet which is observed in the direction of the magnetic force must always originate by symmetric resolution of the original line (§ 69). When I wanted to verify this, I immediately obtained an observation which could not be made to harmonise with a symmetrical situation.

Looking at the doublets of the lines 5791 and 5770, which were very brilliant, I observed a narrow and extremely weak line between the components of the two lines. This weak line seemed, with 5770, precisely midway between the components; with 5791, however, it seemed to be displaced somewhat towards the red."

These weak lines are evidently due to reflection of light, radiating nearly at right angles to the direction of the magnetic force, from the inner surface of the capillary of the Geissler tube. Investigating the neon lines, Lohmann observed a similar perturbation, which in his case was entirely symmetrical. As was to be expected, I found the weak line to be linearly polarised. The whole image, apart from the ratio of intensities and the character of the polarisation, strikingly resembles the type of effect observed at right angles to the magnetic force.

This observation for the line 5791 presented me with the following alternatives: either the resolution of the doublet is dissymmetric with respect to the original line, or the original line is shifted towards the red under the influence of the magnetic field. The latter seemed more probable. Accordingly, I concluded the communication in the meeting of the Amsterdam Academy of Sciences of February 29th, 1908, with these words:

"These experiments forcibly suggest the question:
Has the middle line of a triplet the same wave-length as the unmodified line?

The change of wave-length here contemplated undoubtedly must be extremely small, for none of the physicists occupied with the radiation phenomena in a magnetic field has, to my knowledge, come across phenomena which decide the question put above this paragraph.

Some observations made with an echelon spectroscope have given me evidence that different spectrum lines, and among these the mercury lines, undergo in very strong fields displacements of the order of 6 or 10 thousandth parts of an Ångström unit, in most cases towards the violet. The matter seems of sufficient interest to be treated in a separate paper, which I hope to give shortly."

78. The first part of a paper, "Change of Wave-length of the Middle Line of Triplets," was communicated at the same meeting of the Academy.[1]

More distinctly than with Rowland's grating, a displacement of the central line could be exhibited by means of Michelson's echelon. The observations of which I propose to give an account here have, therefore, been carried out with this apparatus. The photograph, which will be discussed, was published with the second part of my paper. I mention this fact because between the two parts of my paper there appeared one by Gmelin in the *Physikalische Zeitschrift* for April 1st, 1908.[2] Independently of me, and in a different way, Gmelin has treated this question there, and, moreover, has arrived from his measurements at the conclusion that the change of wave-length of the central component is proportional to the square of the magnetic force.

To demonstrate the displacement, I have used the non-uniform field method (see § 71), the observations being made by means of an echelon spectroscope. The displacement must then present itself to the view as a curvature of the central line. It is impossible to observe this curvature with Rowland's grating. The visibility of the curvature is much increased by taking care that, in the image, points corresponding to very different intensities of field lie close together. In order to attain this, an image eleven times reduced of the vacuum tube, charged with mercury and placed into the field, was projected on the slit of the auxiliary spectroscope. The lens used was a photographic objective of 10 cm. focus. Fig. 54 (Plate VI) gives somewhat enlarged reproductions relating to line 5791 and line 5770 respectively.

1 Published in Dutch, March 12th, 1908; in English, March 29th, 1908.

2 Gmelin, "Über die unsymmetrische Zerlegung der gelben Quecksilberlinie 5790 in magnetische Felde," *Physikal. Zeitsch.* (Paper received February 24th, 1908).

To understand these photographs we have only to look again at Fig. 52 and to bear in mind that different orders can be seen simultaneously with an echelon spectroscope. Strictly speaking, only two consecutive orders have a sufficient intensity with an echelon spectroscope.

The lines which bound the image in Fig. 54 for 5770 and 5791 are the central lines of the triplets in consecutive order. On the outside of these lines the outer components curved by the non-uniform field may be imagined, just as in the preceding figure. What lies on the right hand of one central line, and what lies on the left hand of the central line of the other order, will then disappear in consequence of the property of the echelon to which we have directed attention.

The component towards the red in Fig. 54 is always on the left of its middle line, being concave to it in the central part; the second manifestly curved line is the component towards the violet belonging to the other order. The curvature of the middle lines, the demonstration of which is the object of our present experiment, is undoubtedly visible in the figure for 5791. It is still more easily seen by comparison with a straight strip of paper.

In the figure for 5770 this curvature is absent. The dissymmetry of the magnetic resolution of line 5791 is at once evident by the fact that one of the middle lines is approached more nearly by the outer component than the other. The two negatives were taken with the same field intensity of about 34,000 gauss.

It is now abundantly proved that there is a movement of the middle line of a triplet. It is very remarkable that, under the action of the field, the outer components move symmetrically to the original line.

79. For further investigation of the law of movement of the central line, a method warranting greater resolving power than Rowland's grating is needed. Gmelin used Michelson's echelon grating, and it seems that he has largely succeeded by systematic procedure in interpreting quantitatively the results obtained by this instrument. His result therefore possesses great probability.

However, I thought it worth while to investigate the matter by a method independent of Rowland's and Michelson's apparatus. Fabry and Perot's method seemed most appropriate. The greater part of my measurements have been obtained with a 5 mm. étalon, already used on a former occasion. Some determinations were made with an étalon with distance-pieces of *invar*, as suggested by Fabry and Perot, in order to diminish the dependence upon temperature.

The thickness of the air-layer in this étalon was nearly 25 mm. With this distance, and using the light of the mercury line 5791 in the magnetic field, the limit of the method is being rapidly approached. Hence the accuracy of the results obtained with the 25 mm. étalon is in our case scarcely superior to that to be reached with the 5 mm. apparatus.

The arrangement of the apparatus was described with sufficient detail in §§ 9 and 10. For the purpose now in view it was desirable to investigate exclusively the vibrations parallel to the magnetic force. A calcspar-rhomb was placed therefore between the source of light and the first lens. Two images of the radiating vacuum-tube are now obtained near together on the étalon, the undesired image being screened off. A photograph was taken with the field on, and, both before and afterwards, one was taken with the field off. Besides the inner ring, the second ring always, and in some cases also the third and fourth rings, were measured and the result used in the wave-length calculation.

The formula for the calculation is that first given by Fabry and Perot, but greatly simplified in our case (§ 73).

In the following table are given the results relating to

the mercury line 5791. The first column contains the number of the experiment, the second the reference number of the spectrogram; $\Delta\lambda_0$ is the change of wave-length of the central component. The field intensities are given in the last column. Their *relative* values, which are only necessary for establishing the law connecting displacement and strength of field, are exact. These numbers must be increased by 1 or 2 per cent. in order to reduce them to gausses.

Experiment	Plate No.	$\Delta\lambda_0$ in A.U.	H.
1	208^c	0·0085	12700
2	209^b	0·0088	12700
3	211	0·0169	20700
4	212^c	0·0074	13950
5	214^c	0·0201	20600
6	218^b	0·0367	28250
7	218^d	0·0358	28250
8	219^b	0·0360	28250
9	220^b	0·0353	29170
10	220^d	0·0406	29780

Experiments 4 and 5 are made with the 25 mm. étalon; the others with the 5 mm. apparatus. The smallness of the displacements may be illustrated by the statement that the outer components of the triplet 5791 are separated 0·500 Å.U. from the unmodified position in a field of 29,750 gauss. The results 1, 2, and 4; 3 and 5; 6, 7, 8, 9, 10, were combined in each case by assigning simply to each mean displacement the mean magnetic intensity.

Inspection of a graphical representation of the results, or a simple calculation, easily shows that the quadratic law is obeyed within the limits of the errors of observation of the measured displacements. The magnitude of the displacement has been measured in the average in each of the ten points to within 0·002 or 0·003 Å.U.[1]

80. From the displacement of the central line of the triplet it follows, however, that the dissymmetry found in

[1] For further confirmatory experiments, see M. Risco, "La asimetria de los tripletes de Zeeman," *Anales Espanola di Fisica y Chimica*, 263, 1911.

our experiments cannot be explained by Voigt's theory, which suggested the investigation. In most cases the dissymmetries predicted by Voigt are probably too weak to be observed. Efforts have been made to explain the causes of the actually observed dissymmetries, both for the yellow mercury line and in other cases which will be discussed later.

Voigt has advanced an explanation, in which couplings between the electrons are assumed, § 86 below. ("Magneto- und Elektrooptik," p. 261.) It has been pointed out by Lorentz and Dufour that it is possible that in magnetic fields the luminous particles are subjected to a change of structure, which is compatible with the symmetry of the field. Dufour particularly has demonstrated that on this supposition, and for triplets, the displacements of the lines of symmetry of the components with respect to the original line varies directly with the square of the field intensity.

Instead of referring to a change of structure, we may say that the magnetic field modifies the quasi-elastic force in the two principal directions of observation, and that it does so in a different degree.

81. Very striking dissymmetries in the magnetic resolution of absorption lines of solid substances have come to light in the investigations of J. Becquerel in Paris, of H. du Bois and G. J. Elias in the Bosscha Laboratory in Berlin. J. Becquerel investigated the absorption bands of some crystals containing rare earths: xenotime, tysonite, and parisite. Moreover, the influence of the temperature was studied for these spectra; first in liquid air, and then by Kamerlingh Onnes and J. Becquerel in the Leyden Laboratory, at temperatures down to $-260°$ in liquid and solid hydrogen and in fields of 18,000 gauss. At the ordinary temperature and in liquid air, du Bois and Elias investigated several chromic compounds (ruby), a number of salts of the rare earths; also, bastnaesite and hussakite in fields up to

40,000 gauss. The three first-mentioned crystals and hussakite, erbium and yttrium sulphate, and erbium nitrate offer a great number of examples of dissymmetries with respect to place and intensity. In still closer connection with the above proved displacement towards the red of the central line of the triplet of the mercury line 5791 is the fact observed by J. Becquerel[1] that in the absorption spectrum of xenotime the centre of the doublet of the line 5221 is moved over a distance of 0·1 to 0·2 Å.U. towards the violet in a field of 24,000 gauss.

In 1910 observations of the line spectrum of chromium were published by Dufour.[2] For several chromium lines the symmetrical line of the vibrations as well of those parallel to the field and those at right angles to it, are moved towards the violet under the influence of the magnetic force. The amount is different for different lines, and sometimes reaches 0·08 of the value of the normal resolution. (For components which undergo a normal resolution the value of the ratio $\delta\lambda/H\lambda^2$ is 0.94×10^{-4}, according to the measurements of Cotton and Weiss and Gmelin. In this λ is expressed in cm., H in gauss; $\delta\lambda$ is the distance of the outer components of a triplet.) *Cf.* p. 67.

82. I will borrow an example of a special dissymmetry from the extensive investigations of J. E. Purvis, which refer to a number of elements. The chromium line 3209.3 is split up into a quintet. Three components (viz. that at the place of the original line, and the two outer) vibrate perpendicular to the field, and two parallel to the field. The dissymmetry of the intensities is now so great that on the photographic plate the most refrangible outer component fails almost entirely, so that the whole gives the impression of a quadruplet.

Near the mercury line 5790.87, the peculiar behaviour of which in the magnetic field we have fully discussed, a

[1] J. Becquerel, *Compt. rend.*, **148**, 914, 1909.
[2] Dufour, *Journ. de Phys.*, April, 1910.

satellite is found on the side of the blue with the wave-length 5789.88. According to Gmelin (*Physikalische Zeitschrift*, **11**, 1193, 1910) this line moves towards the violet in direct ratio to the field intensity, splitting up into a symmetrical triplet, the middle component of which is the most intense. The polarisation of this triplet is remarkable; for though the middle component is un-polarised, the component towards the violet vibrates perpendicular to, that towards the red parallel to, the magnetic force.

83. Our own experiments on mercury and those of Gmelin were made by means of vacuum tubes; those of Purvis with electric sparks. In these cases strong electric forces act conjointly with the magnetic field, and one would be somewhat inclined to ascribe the observed anomalies to them. In connection with a theory of Voigt on an electric analogue of magnetic resolution, I have proved that some observed anomalies[1] might be explained by a co-operation of magnetic and electric forces, if only the intensity of the electric forces in the spark or in the vacuum tube be taken sufficiently great. Two courses may be followed to inquire into this point.

In the first place, observations may be made with flame spectra to see if the dissymmetries then disappear, since there are no electric, external fields in the flames.

In the second place, we may make use of the theorem proved by me that a dissymmetry which would be the consequence of a co-operation of magnetic and electric forces reverses its sign, if the direction of the electric field is rotated through an angle of 90°. For example, if the electric force is parallel to the magnetic force, the distance between the middle component and the outer component towards the red might be smaller than that between the

[1] Cf. Zeeman, "Considerations concerning Light Radiation under the simultaneous influence of Electric and Magnetic Forces and some experiments thereby suggested," *Proc. Acad. Amsterdam*, May, 1911.

middle component and the outer component towards the violet. This relation is reversed when a horizontal magnetic field is combined with a vertical electric one.

84. To decide the first question, whether dissymmetries occur also in flame spectra, in conjunction with Dr. Winawer, I made some experiments [1] on this subject. We examined some lines of cæsium, barium, and tin, which were obtained by the introduction of filter paper soaked with solutions of the salts of these elements into a gas-hydrogen flame. With a small echelon spectroscope, some dissymmetrical resolutions could be observed.

The most cogent proof of the existence of dissymmetries in flame spectra, however, is furnished by the investigations by Dufour on chromium lines already cited. Dufour made these observations with an oxygen-acetylene flame, with which he obtained lines which were more delicate and intense than when he worked with a condensed spark and self-induction.

Finally, in vindication of the view that no external electrical field is required to obtain dissymmetries, we may refer to the cases studied by J. Becquerel, du Bois and Elias, in which there was no question of an exterior electric field.

85. I made some observations [2] with the yellow mercury lines, which were suggested by the second view given in § 83. As we have already in this chapter repeatedly had occasion to observe: in a horizontal, magnetic field in which a vertical vacuum tube is placed, and in which we may therefore assume vertical electrical forces, the mercury line 5791 exhibits a dissymmetrical triplet. The distance of the outer component towards the red from the central one is distinctly smaller than that towards the violet.

A new experiment was made with a du Bois electromagnet. The axis of the magnet was adjusted vertically, and a mercury tube of a particular shape was introduced in the axis of the magnet through the pierced cores. Then part of the luminous vacuum tube remained visible. In this part, magnetic force and electric force were parallel. An image of the portion of the tube was projected on the slit of the spectroscope. The dissymmetry appeared to be exactly the same as had been observed before with electric and magnetic forces at right angles to each other. Hence we may conclude that in all the cases mentioned the magnetic force is solely responsible for the dissymmetry.

86. The results of this chapter may be thus summarised. Some emission lines of the line spectra, a few emission bands, and several absorption lines of solid substances exhibit magnetic resolutions which are dissymmetrical in place and intensity. Parallel and at right angles to the field the dissymmetries are the same. Some spectrum lines are shifted under the action of the field, the mean of the wave-lengths of the components being changed. These dissymmetries present many difficulties to theory. Their experimental investigation, on account of the minuteness of these sub-effects, is not exempt from difficulties.

The cause of the dissymmetries will have to be found either in couplings between electrons with different frequency, or in a change of structure of the radiating particles under the influence of the magnetic field. When interpreted in the light of the first-mentioned hypothesis, a shift as observed for the mercury line 5791 must be due to a coupling with another line in its vicinity. Very recently Paschen and Back have published remarkable experiments (see § 121) which are essentially consequences of the shift-phenomenon. In the experiments of these physicists the interacting lines are fully known in their behaviour when separate. The advantage of the theory of coupled electrons in the theoretical interpretation of the observed phenomena seems manifest.

CHAPTER VIII

SOLAR MAGNETO-OPTICS

87. IN discoveries of optics we may always cherish the hope that they will lead eventually to applications to astronomy. Shortly after the discovery of spectrum analysis, studies of the light of stars suggested the unity of the chemical constitution of the universe; and in the same way Doppler-Fizeau's principle has been widely applied in astronomy. The magnetic resolution of the spectrum lines was required by G. E. Hale for the brilliant discovery of the magnetic fields at some points of the sun. We will devote this chapter to Hale's discovery. Before examining in detail what characteristic phenomena of magnetic resolution are found in sun-spot spectra, some general remarks on sun-spots may be desirable.

A direct photograph of a sun-spot or a group of sun-spots generally presents but few remarkable peculiarities. The case is different, however, when photographs are taken with a spectroheliograph, by means of which an image of the sun's surface is formed in the colour of a single spectrum line. In the hands of Deslandres and Hale this method has greatly extended our knowledge of the sun.

The whole surface of the sun displays a detailed structure (Fig. 55). A photograph taken with the light of one of the calcium lines is entirely different from the directly observed image. If the light from higher layers of the sun is used, and if, like Hale,[1] the red hydrogen line Hα is selected, new unforeseen possibilities of research are revealed. In a great many cases Hale's beautiful photographs point to cyclonic movements about the spots, in which matter is whirled round in a "solar vortex" of gigantic dimensions (Fig. 56). The axis of the vortex is approximately in the direction of the radius of the sun,

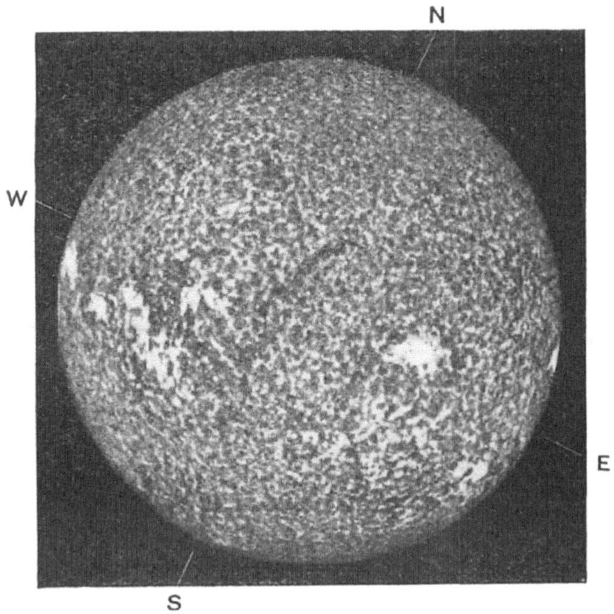

FIG. 55.—SPECTROHELIOGRAPH PLATE TAKEN WITH VIOLET SIDE OF K_2, SEPTEMBER 18TH, 1908.—DESLANDRES.

and coincides with the centre of the spot. As the solar gases undoubtedly contain free electrons, Hale was led to inquire whether these electrons would be whirled round with the vortex. According to known laws, these moving electrons will then produce a magnetic field, and provided this be sufficiently strong, the spectrum of the

[1] Hale, "Solar Vortices," Contributions to Mount Wilson Sol. Obs., No. 26.

sun-spots would give an optical manifestation of this magnetic field.

88. The spectrum of a sun-spot was first observed by Sir Norman Lockyer in 1866.[1] He discovered that many dark lines of the solar spectrum are widened over the full extent of the spot. Since then these widened lines have been systematically examined by numerous observers. If the observations are made with an instrument of high dispersion, it will be seen that some of the iron lines

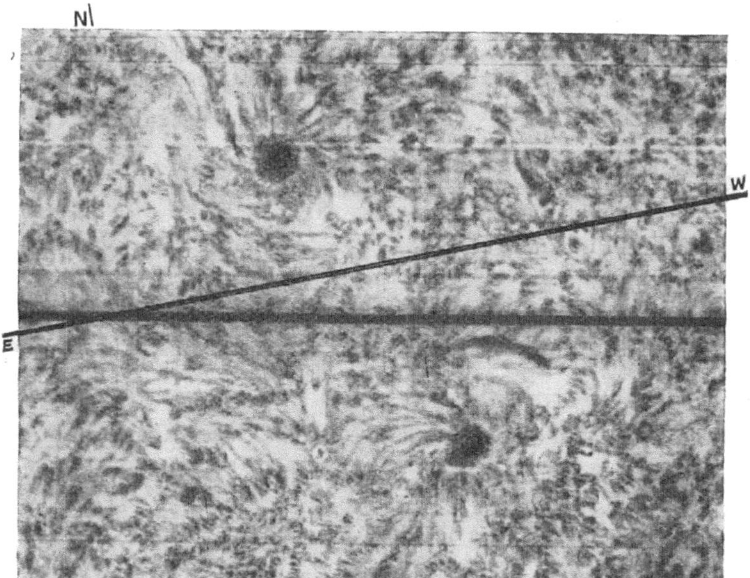

FIG. 56.—SUN-SPOTS AND HYDROGEN FLOCCULI, SHOWING RIGHT- AND LEFT-HANDED VORTICES, OCTOBER 7TH, 1908.—HALE.

which are single in the solar spectrum are double in the spot spectrum. Such double lines were first seen by Young, of the Princeton Observatory, with a large spectroscope attached to a 23-inch refractor.

W. M. Mitchell,[2] who worked with Young for some time, gave a drawing of different types which are met

[1] Norman Lockyer, *Proceedings R. S.* **15**, 256, October, 1866.
[2] W. M. Mitchell, *Astrophysical Journal*, **22**, 4, 1905.

with in the spot lines. This drawing is reproduced in Fig. 57 (Plate VII) in a somewhat modified form.[1] Mitchell subjected the chromospheric lines and the lines "affected" in the spot-spectrum to an exhaustive comparison. He described the doublets as "reversed lines," which they, indeed, closely resemble. It seemed that they might be considered as a bright line at the centre of a widened dark line. A brightly radiating layer of vapours lying above the spot would then have produced the narrow bright line.

A widening and a doubling of the lines would, however, also have to be expected if a sun-spot in the centre of the sun's disc may be considered as a solar vortex with a magnetic field directed along the sun's radius. The characteristic properties of the inverse longitudinal spectrum effect would then have to manifest themselves. Led by this train of reasoning, Hale resolved to test the components of the spot doublets for evidence of circular polarisation and to seek for other indications of the magnetic separation.

89. The "Tower" telescope of the Mount Wilson Solar Observatory forms an image of the sun about 6·7 inches (17 cm.) in diameter in a horizontal plane three feet above the surface of the ground on the slit of a spectroscope. The beam of sunlight, after passing through the slit, descends vertically into a well about thirty feet deep, excavated in the earth. Thirty feet from the slit the diverging rays encounter a 6 inch (15 cm.) objective, through which they pass in a parallel beam. After diffraction by a 4-inch plane Rowland grating having 14,438 lines to the inch, the rays return through the objective. An image of the spectrum is made to fall at a point slightly to one side of the slit; here it may be recorded photographically. A Fresnel rhomb and a nicol are mounted in front of the slit. In addition to each spot spectrum, the

[1] *Trans.* International Un. Sol. Res., **2**, 204, 1908.

Fig. 57.

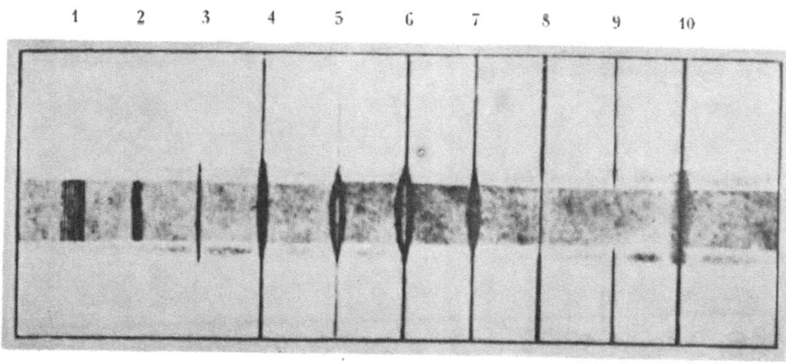

Types of sun-spot lines. (Mitchell)

5, 6. Widened Lines with Centres reversed Bright.
7. Widened and Weakened Line. 10. Winged Line.

Fig. 58.

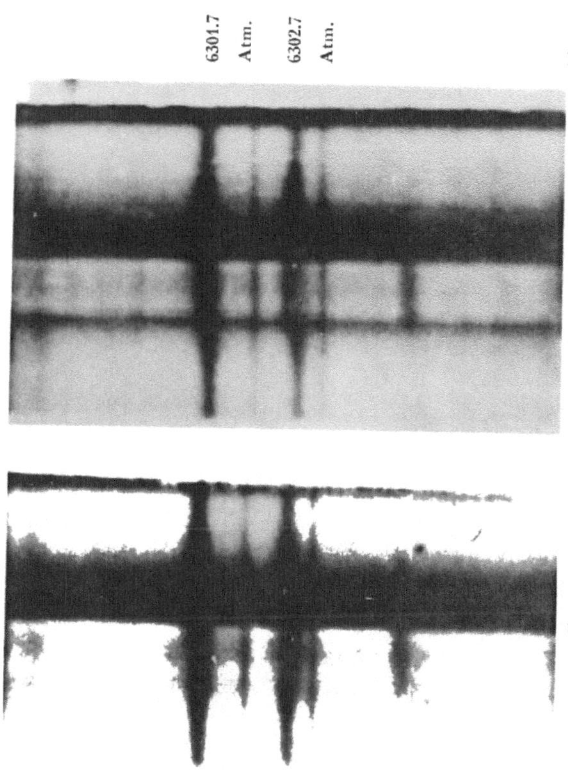

spectrum of the photosphere on either side of the spot can be photographed for comparison.

Photographs were taken with different positions of the nicol covering a range of 90°. When a true magnetic doublet is present, then in a certain position of the nicol, the light from the red component should be transmitted and that of the other component cut off.

Because the light from the spot does not travel, in general, parallel to the magnetic lines of force, the doublet may exhibit traces of elliptical polarisation and the extinction of either component may be incomplete. The first result of this test may be quoted in Hale's[1] own words :—

"My first observations were made on June 24th [1908], in the second order of the grating, but the results were not conclusive. On June 25th I obtained some good photographs, in the third order, of the region λ 6000—6200, using Seed's 'Process' plates, sensitised for the red by Wallace's three-dye formula. These clearly showed a reversal of the relative intensities of the components of spot doublets, when the nicol was turned through an angle of 90°. Moreover, many of the widened lines were shifted in position by rotation of the nicol, indicating that light from the edges of these lines is circularly polarised in opposite directions. The displacements of the widened lines appeared to be precisely similar in character to those detected by Zeeman in his first observations of radiation in a magnetic field."

90. We quote here from a paper that appeared in the November (1908) number of the *Astrophysical Journal*, but perhaps a more historical treatment may be given with advantage. I may first mention a letter, dated, July 6th, 1908, which Hale very kindly sent to me, accompanied by a copy of a manuscript of a note on

[1] Hale, Contributions Mount Wilson, No. 30, p. 7, *Astrophysical Journal*, Nov., 1908.

"Solar Vortices and the Zeeman Effect."[1] I further received two photographs on glass of double lines in the spot spectrum between two comparison spectra of penumbra and photosphere of the region λ 6250—λ 6360. These were taken with the arrangement of nicol and rhomb already mentioned, the position of the nicol being changed between the first and the second of these photographs. The accompanying illustration (Fig. 58, Plate VIII.) is from an enlargement of this photograph and relates to the iron line at λ 6302.71. It will be seen that the relative intensities of the two components, or more exactly of the edges of the widened line, were reversed by turning the nicol through an angle of between 45° and 90°. The narrow lines on both sides of the line 6302.71 are of atmospheric origin. Hale asked me to examine his photographs, and to send a note to *Nature* expressing my opinion as to the interpretation of the results. I did so,[2] and the writing of my note for *Nature* was a delightful incident of my holidays.

There could be no doubt that Hale gave what appeared to be decisive evidence that sun-spots are strong magnetic fields, the direction of these fields being nearly perpendicular to the sun's surface.

Indeed, the phenomena observed in the double lines and the widened lines of the sun-spot spectrum, and exemplified in the photographs under review, are identical in character with those observed with the longitudinal inverse effect as given in Chapter II. of this volume. Only a strong magnetic field can resolve a single line into a doublet with components circularly polarised in opposite directions. Whenever these characteristic and unique phenomena are present, a magnetic field must be their cause.

The absence of any shift of the red telluric lines by the

[1] A brief abstract of this note appeared in *Nature* for August 20th, 1908, together with a short paper by the present writer. Hale's note was printed in the Publications of the Astronomical Society of the Pacific, **20**, 220, 1908.

[2] *Nature*, **78**, 369, 1908.

rotation of the nicol, or of measurable displacement of the cyanogen flutings, considerably strengthens the argument and excludes instrumental and other errors. A quantitative comparison of the magnetic separations of the iron lines as observed in the laboratory and in the sun could not be made at the time, as the separations of the red region of the spectrum were not available. The iron spectrum in the magnetic field had been investigated fairly carefully, but the lines examined were not far enough in the red.

An estimate of the intensity of the sun's magnetic field could be hazarded by using the fact given by Hale that the separation of many doublets ranges from 0·018 to 0·216 Å.U. The latter number would give, on the supposition of a high specific separation of the line under review, a field intensity of the order of 6000 gauss.

A magnetic field of the observed order of magnitude might conceivably be produced in the solar vortex without greater crowding of electrons than is to be found in an ordinary vacuum tube.[1]

91. The first observations of Hale refer, accordingly, to the circular polarisation for the longitudinal effect. The sun-spot was observed when it was near the centre of the sun's disc. The question may be asked whether it is also possible to observe the transversal effect, in which case the lines must then have the appearance of triplets with linear polarisation. In this observation the spot must be on the sun's limb.

Another question which suggests itself, is whether the sign of the circular polarisation is reversed when the direction of rotation in the vortex reverses.

The answer to both questions was given in a telegram from Hale, which I received on September 21st, 1908, and had the privilege of reading before the meeting of the German "Naturforscher und Ärztë," which was then assembled at Cologne, on the 23rd of the same month.

[1] Zeeman, *Nature*, *l.c.*

The telegram, as short as important, ran: "Vortices rotating opposite directions show opposite polarities; spot lines near limb plane polarised." So the transversal effect had also been observed in the spots.

92. With the aid of Hale's exhaustive paper, we will make a few additional remarks to supplement what has been said. In the first place we shall state something more about the comparison of the amount of the resolution in the spots with the analogous resolution found in the laboratory. Many lines in the spark-spectrum being very faint, this comparison is not without difficulty.

In the second column of the following table are given, under the heading "Δλ, Spark," the distances of the outer components as measured in the laboratory in a field of 15,000 gauss, the third column contains the quantity given in column 2 divided by 5·1; the fourth the separation of the components observed in the spot-spectrum; and in the last the ratio of the separation in the spark to the observed separation in the spot.

Iron Doublets.

Wave-length.	Δλ, Spark.	$\frac{\Delta\lambda, \text{Spark}}{5{\cdot}1}$.	Δλ, Spot.	$\frac{\Delta\lambda, \text{Spark}}{\Delta\lambda, \text{Spot}}$.
6213·14	0·703	0·138	0·136	5·2
6301·72	0·737	0·144	0·138	5·3
6302·71	1·230	0·241	0·252	4·9
6337·05	0·895	0·175	0·172	5·2

The agreement is so close that it cannot be the result of chance.

The intensity in the field of the electromagnet being 15,000 gauss, the intensity of the field in the spot must be about 15,000 divided by 5·1, or 2,900 gauss. The most intense fields measured by Hale are 4,500 gauss.

For the investigated titanium and chromium lines, the agreement between the results for the spots and for the spark is less satisfactory. The two D-lines and the *b*-lines of magnesium are but slightly affected in the spot-spectrum. As the elements magnesium and sodium probably occur at a much higher level in the spots, the conclusion seems justified that the intensity of the magnetic force in the spots rapidly decreases with the height.

93. In many cases it appeared that the normal spectrum of the spots contained both triplets and doublets. As this is also the case for spots near the centre of the solar disc, the conclusion follows that the light is emitted by the spot in a direction which makes a considerable angle with the direction of the magnetic force. The observations are made obliquely with respect to the magnetic force. The amount of this angle may be approximately determined from the ratio of the intensities of the lateral components to the central component of a normal triplet. In a direction at right angles to the magnetic force, the intensity of the middle component must thus be twice that of a lateral component, the intensity of the middle component becoming less as one gets nearer to the direction of the force. It varies in proportion to the square of the sine of the angle formed by the ray of light and the magnetic force.

The projection of the magnetic force on the solar disc can also be directly found. It is parallel to the plane of vibration of the nicol in the position in which the middle component reaches its maximum of darkness.

In this way it would be possible to make a map of the magnetic fields on the sun. The direction of the magnetic force in space would be entirely determined, and also the absolute value of the force, since it is given by the magnetic separation.

On account of the fact that only few spectrum lines behave as normal triplets, we are confronted by a difficulty of a different nature in the determination of the exact direction of the magnetic force. We shall return to this subject in the next chapter. It already appears, however, with perfect certainty from Hale's later investigations that the magnetic axis of a vortex does not, in general, coincide with the solar radius.

This inclination of the axis of the vortices seems to be in favour of Emden's solar theory. This theory recognises the fact that two spots on either side of the sun's equator may be related. They may be the extremities of one and the same vortex tube. The rotatory direction of the two spots would have to be different for the two hemispheres. Hale once photographed such a case with the spectroheliograph, and the opposite signs of the magnetic field observed in the two spots is, of course, a forcible argument in favour of Emden's theory.

The rapid decrease of the magnetic field of the sun-spots with the height proves, indeed, that the magnetic action of a spot will not reach far. Moreover, Schuster has expressly pointed out that even if this rapid decrease of the field did not exist, the action of numerous spots with magnetic fields of the order of those found would be altogether inadequate to account for terrestrial magnetic storms.

94. In Hale's first observations the fact that many lines continued to be doublets, even when the sun-spots had approached the sun's limb, seemed a serious argument against the magnetic field hypothesis. It has been shown, however, particularly through King's work, that many seeming doublets in the iron and the titanium spectrum are in reality quadruplets, for which a few components are too close together for resolution in the spot spectrum. Such lines as 6173.55 and 6302.71 are, however, fine triplets.

95. On Mount Wilson the further interesting discovery has been made[1] that vortices are often revealed by the magnetic separation. In some cases what is photographed does not appear to be a vortex, but there must be a vortex present. The electric vortex is the only cause that will

[1] *Proc.* Fourth Intern. Solar Conf., **3**, 23, 1910.

produce the characteristic properties observed in the spectrum of sun-spots.

96. In the spectra of some sun-spots the lines are parallel throughout their entire length. In others the lines converge. As the magnetic separation measures the intensity of the field, these facts only mean that the intensity varies differently in different cases.

In general, the maximum intensity appears in the central part of the umbra, and decreases rapidly across the penumbra of the spot. The lines converge over the penumbra. It is interesting to compare Mitchell's drawing with the photograph relating to the mercury lines in a non-uniform field (see § 87 above and § 71, Fig. 52, also § 111).

Quite recently Hale[1] has made an attempt to observe manifestations of the weak general magnetic field of the sun. The quiet condition of the sun, *i.e.*, the absence of sun-spots during the year 1912, made it possible to look for such small magnetic effects. Experiments were arranged to detect displacements of the lines due to extinction of their elliptically polarised red or violet edges by a compound quarter-wave plate (*cf.* § 65) and by the nicol. Some of the solar lines gave shifts as great as 0·0024 Ångström. Photographs were taken north and south of the equator and at different latitudes. The values of the displacements decreased from a maximum near 45° to about zero near the equator, and they were of opposite sign in the two hemispheres. This should be the case for a line which behaves in a normal way, as seen by an observer in the plane of the sun's equator. The magnetic separation of the lines observed in the sun could not yet be measured in the laboratory on account of their extreme faintness in the spark.

[1] *Hale*, "Preliminary Note on an Attempt to Detect the General Magnetic Field of the Sun," *Terrestrial Magnetism*, **17**, 173, 1912.

CHAPTER IX

THE INVERSE EFFECT IN DIRECTIONS INCLINED TO THE FIELD. APPLICATIONS TO SUN-SPOT SPECTRA.

97. The inverse effect, *i.e.*, the magnetic separation of absorption lines, apart from its physical interest, has become of great importance in solar physics since Hale's discovery of the corresponding phenomena in sun-spot spectra. We have hitherto limited our consideration to the two principal directions (parallel and at right angles to the lines of force); in this chapter we shall consider the phenomena which result when the observations are made obliquely to the lines of force—a case which becomes especially interesting, if the components of the line under magnetic influence are not neatly separated. The laws of this general case apply directly to solar magneto-optics.

Let us begin with the separation of a narrow spectrum line into a triplet. According to the elementary theory of magnetic division, the total radiation of an electron vibrating in a magnetic field is made up of three parts. Two parts correspond to circular vibrations in planes perpendicular to the magnetic field, taking place in opposite directions, and a third to linear vibrations parallel to the field.

If we examine the light in one of the principal directions, we arrive at results which have been described in this volume more than once. It is easily proved that, in a direction oblique to the lines of force, a triplet with elliptically polarised outer components may be observed.

The ellipse, which characterises the state of polarisation of the components with a period $T+\delta T$, is the projection on the wave-front of the circle perpendicular to the field in which the electron with period $T+\delta T$ is supposed to be moving. The direction of the motion of the moving electron also determines the motion in the ellipse. The ratio of the axes of the ellipse is as 1 to cos θ, θ being the angle between the direction of the field and the ray. For the other outer component with period $T-\delta T$ the same reasoning holds, *mutatis mutandis.* The central line with the unmodified period T always remains linearly polarised. The vibrations of the middle component are in the plane determined by the ray and the line of force and the amplitude of the vibrations is proportional to sin θ.

If we put $\theta=0$, *i.e.*, in the case of the longitudinal effect, only circular motions remain.

All this applies to very narrow spectrum lines in a strong field, the distance of the components being much greater than their width. According to Voigt and Lorentz, we must expect some interesting peculiarities if this restriction be discarded. We will return to this point later. As a general rule, the deductions from the elementary theory are verified. Also, in the case of the quartet and the sextet, the outer components become elliptically polarised, as has been observed already by Righi. In contradiction with the elementary theory, though not strictly applicable to the case, is the very slight diminution of intensity of the middle components of the quartet even for $\theta=45°$.

The ellipticity of the outer components is easily proved by means of the arrangement described in the third chapter, if we introduce in the beam also a Fresnel rhomb with its principal plane at an azimuth of 45° with the horizon. Fig. 59 (Plate VIII) relates to an angle θ of 60°, and is taken from a memoir by Zeeman and Winawer (see § 100 below).

The outer components of the quartet towards the red or

towards the violet, dependent upon the strip and the direction of the field, are now considerably weakened. All this proves the elliptical polarisation of the outer components.

The ellipticity may be proved also by the calc-spar rhomb alone: at least in some cases. In others the use of the Fresnel rhomb is more efficient. We must refer to the complete memoir for a discussion of this point.

98. Let us now discard the supposition that we have to do with *very narrow* spectrum lines observed under the influence of a *strong* field. The more general theory covering also the case of magnetically separated lines, which partially overlap, has been developed by Voigt and by Lorentz. In Voigt's considerations, it is the inverse effect that is considered in the first place. Conclusions for the emission in magnetic fields may be drawn by referring to the classical connection between emission and absorption. This apparent round-about way is justified by the fact that the problem of the absorption exercised by a system of reciprocally reacting electrons is much easier than the corresponding emission problem. The peculiarities of absorption are represented by the addition of a frictional term to the equations of motion for the propagation of light. The important results obtained by Voigt, consisting in the establishment of a rational connection between the Faraday effect and magnetic separation, and between the last mentioned effect and magnetic double refraction, are discussed in Chapter V. The results obtained by the theory of propagation of light in an absorbing medium submitted to magnetic forces are relatively simple. In order to introduce the subject, we shall consider briefly light propagation in ordinary double refracting media.

In a crystalline medium such as Iceland spar, plane waves may be propagated; the vibrations in each plane are linear and parallel to two possible directions. The velocity of propagation depends upon the direction of the vibration. There are, therefore, in every direction in the crystal two *principal beams*, each characterised by its own direction of vibration and corresponding velocity of propagation. The directions of vibration in the two beams are at right angles to each other. The absorption indices of the two are equal.

This is no longer the case if we have to do with an absorbing crystal; the principal beams then cease to be linearly polarised; they are elliptically polarised.

In the case of solutions of copper tartrate and chromium tartrate in potash, Cotton found (*cf.* § 43) a rotation of the plane of polarisation (in the absence of a magnetic field) and a different absorbing power for oppositely circularly polarised vibrations. Ordinary unpolarised light is decomposed in the solution in two oppositely circularly polarised beams, with different velocities and different absorption indices.

A close analogy with the behaviour of these solutions is presented according to theory by the magneto-optical phenomena. For arbitrarily chosen values of the angle θ between the ray and the magnetic force for every frequency, two principal beams with elliptical vibrations of opposite directions can be transmitted through the magnetised metallic vapour. The velocities of propagation and the absorptions of the two beams are different and depend upon the angle θ.

We shall now summarise briefly the chief results of Lorentz's[1] considerations concerning our present subject.

If we are not dealing with a sharp triplet, *i.e.*, three absorption bands that are completely separated, we can still say something about the vibration-ellipses of the outer components.

Let axes OY and OX' be chosen, one normal to the plane passing through the ray and the magnetic force, the

[1] Lorentz, "On the Theory of the Zeeman Effect in a direction inclined to the lines of force," *Proc.* Acad. Amsterdam, **18**, 321, 1909.

other perpendicular to the ray and lying in the plane just mentioned. Then one of the characteristic vibration-ellipses can be considered as the reflected image of the other with respect to a line bisecting the angle $X'OY$ (Fig. 60). This rule also applies to the direction of motion in the two ellipses.

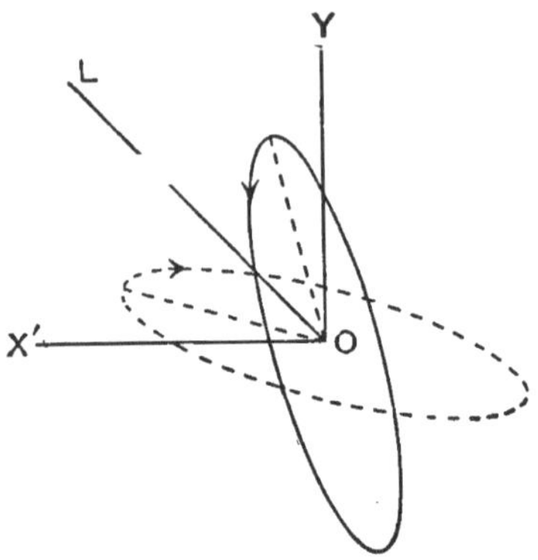

FIG. 60.

As in the present chapter we suppose the lines of force always to be horizontal, we examine the propagation of light also in a horizontal plane. From what has just been said, we may infer therefore that the axes of the vibration ellipses are inclined to the vertical. As the absorption indices of the two elliptically polarised beams differ greatly, we may expect in general that after passage through the vapour the issuing light exhibits elliptical polarisation.

If the vapour becomes very dilute, the results obtained agree with that of the elementary theory; the major axes of the characteristic ellipses coincide with the vertical and horizontal lines.

99. Other predictions of the theory relate to the frequency n_o of the central line. The nature of the phenomena that may be observed for $n=n_o$ depend upon the value of θ being greater or smaller than a certain angle θ_1. This latter is determined by Lorentz's equation

$$\tan \theta_1 \sin \theta_1 = \frac{g}{\nu}.$$

The quantity g may be regarded as a measure of the width of an absorption line and depends upon the constants of the vapour; ν is determined by the change of the frequency of the free vibrations of the electrons and has a value proportional to the strength of the field.

If $\theta > \theta_1$, then two linearly polarised beams with equal indices of refraction and different absorptive indices can be propagated. The rectilinear vibrations make equal angles with the line OL (Fig. 61), bisecting the angle $X'OY$. The absorption is stronger for the beam the vibrations of which make the smaller angle with the direction of the field. In the figure the more strongly absorbed vibration is indicated by a thicker arrow. As θ decreases, the vibrations of the two principal beams approach more and more to OL, so that for $\theta = \theta_1$ both directions coincide with the bisectrix. The two principal beams are now equally absorbed also.

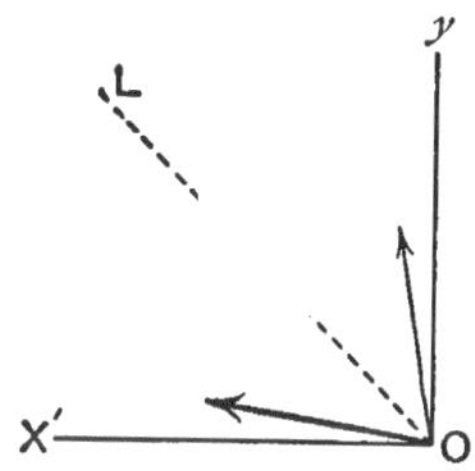

FIG. 61.

When $\theta < \theta_1$, the state of things is wholly different. In this case two elliptically polarised beams can be propagated; they are equally absorbed, but have different velocities of propagation. For these two beams the characteristic ellipses are the same, but described in opposite directions. One of the axes of the ellipses coincides with the line OL in Fig. 61. The ellipses become less and less eccentric as the wave becomes less inclined to the direction of the field. For $\theta = 0$ the ellipses become circles described in opposite directions. A further approximation for $\theta = \theta_1$ shows that in this case the two vibrations do not coincide exactly. As in the general case, there are two distinct beams with different characteristic ellipses, both deviating somewhat from the line OL of Fig. 61. The regions of the longitudinal and the transverse magnetic effect overlap to a certain extent and are not sharply separated from each other at the angle θ_1.

100. There are three results of Lorentz's theory that admit of experimental verification and were verified in an investigation made in collaboration with Dr. Winawer.[1] Let us imagine the absorbing vapour placed in such circumstances that the elementary theory cannot be applied. The components of a divided line are now not neatly separated by practically transparent regions. The vapour density must be chosen relatively great and the magnetic intensity rather small.

The three predictions referred to, which apply, if we exclude the cases of the true longitudinal and transverse effects, are: (1) the major axis of the vibration-ellipses of the outer components deviate from the vertical line; (2) the vibrations of the middle component (or components) are, depending on circumstances, either linear and not horizontal, or elliptic, the axes of the ellipse being inclined to the horizon; (3) there exists an angle θ_1 separating the regions of the longitudinal and the transverse magnetic effect.

Oblique Position of the Vibration Ellipses of the Outer Components.

101. We succeeded in establishing experimentally the oblique position of the vibration-ellipses in the inverse magnetic effect of the D-lines; the amount of the slope of the axes could be measured. The obliquity is far from striking. When θ was already such that the ellipticity was very marked, it was only with some difficulty that we could make sure of the obliquity. Some details of a definite case may be given. With $\theta = 69°$ and a field of about 18,000 gauss, the first observations were made. Attention was given to D_2, the vapour-density being regulated so that the outer components of the sextet could not be seen separately. Before the slit of the spectroscope a nicol was placed with its plane of vibration at an azimuth

[1] P. Zeeman and B. Winawer, "The Magnetic Separation of Absorption Lines in connexion with Sunspot Spectra," *Proc. Acad. Amsterdam*, 1910; *Astrophysical Journal*, Dec., 1910.

of, say, 35° with the horizon. The central part of the resolution figure is now very dark; the outer components of the pseudo-triplet, however, are only faintly visible. This has the advantage of increasing the visibility of small changes of the intensity of the outer components.

The direction of the field we denote as field-direction 1. *With the reversed field-direction 2, the outer components became darker.* The nicol then was placed in a position symmetrical to the one just mentioned. Now with field-direction 1, the outer components were darker. From these experiments we must conclude that a vertical line is not an axis of symmetry of the vibration-ellipses of the

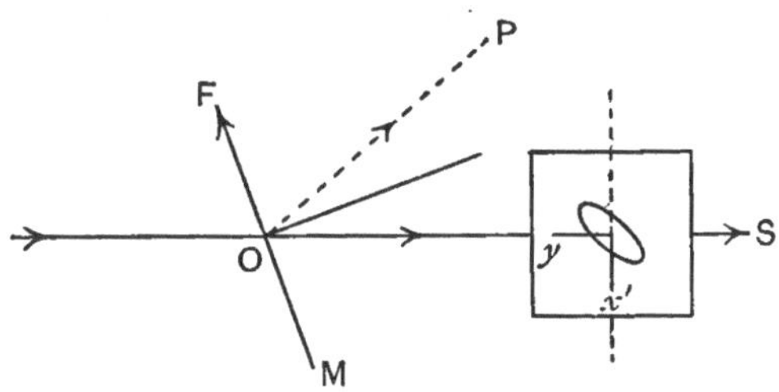

Fig. 62.

outer components, hence that the position of these ellipses is oblique.

102. The direction of the smaller axis of the vibration-ellipse we measured for $\theta = 69°$, the vapour-density being between the first and second phase (*cf.* § 33). In front of the slit of the spectroscope was introduced a nicol, mounted upon a divided circle which gives the rotation of the nicol in degrees. The vanishing or reappearing of the outer components gave a good criterion for the determination of the smaller axis and therefore of the major axis of the vibration-ellipse. The measurements gave the result that in the circumstances of the experiment the major axis made an angle of 5 degrees with the vertical. The obliquity was the same in amount and direction for the components toward the red and toward the violet. The diagram, Fig. 62, illustrates the relation between the slope of the ellipses and the direction of the field.

Let OS be the beam, which traverses the source of light placed in O, and OF the direction of the magnetic force. For an observer looking in the direction SO, the upper part of the vibration-ellipse is inclined toward the right. The plane yx', containing the ellipse, is normal to the ray and in the figure has been rotated round the dotted line until brought into coincidence with the plane SOM. That side of the plane which was visible from S can now be seen. Both the ellipse toward the red, and the ellipse described in the opposite direction toward the violet, have the same slope with a given direction of the magnetic field, as was remarked above.

103. *The same* sodium flame, investigated as to the inverse effect in the direction OS, we studied in the direction OP (*i.e.*, for an angle $FOP = MOS = 180° - \theta$) for the phenomenon of partial polarisation, discovered by Egoroff and Georgiewsky. A small telescope focussed upon the flame was used and provided with a Savart plate and a nicol. This polariscope is mounted upon a divided circle graduated in degrees. The direction in which the fringes were most brilliant was determined in order to detect a possible deviation of the plane of maximum polarisation from the vertical. It was easily seen that there was such a deviation. The fringes were clearest if for the observer in P their direction was from the upper left to the lower right quadrant, the direction of the field being always as indicated in the figure. After reversal of the magnetic field the fringes become indistinct. They become distinct again if the principal direction of the polariscope was from the upper right to the lower left quadrant. The result of these observations at least proves that the whole phenomenon is asymmetrical with respect to the vertical, and hence proves the presence of oblique vibrations. The observation with the Savart

(60°)

Fig. 59.

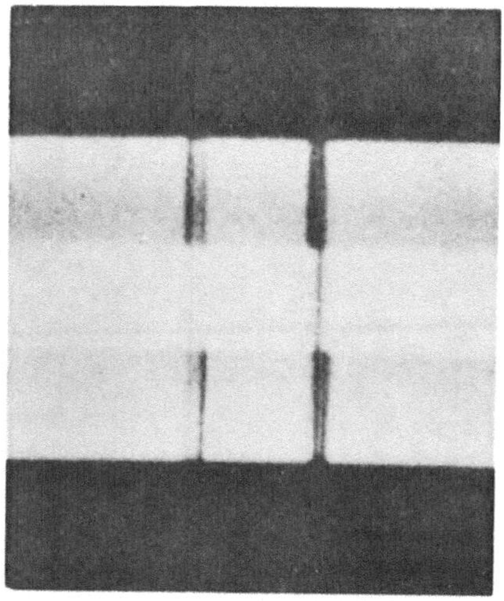

D_1 D_2

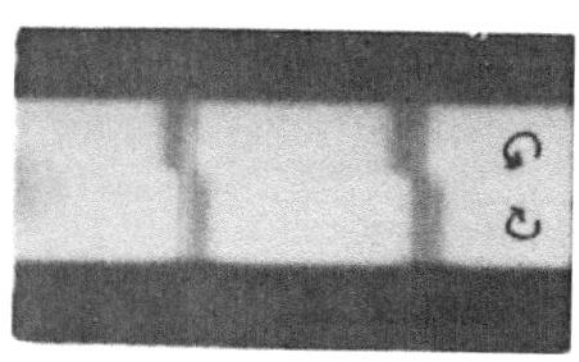

Fig. 67.

5^1 6^1 7^1 10^1

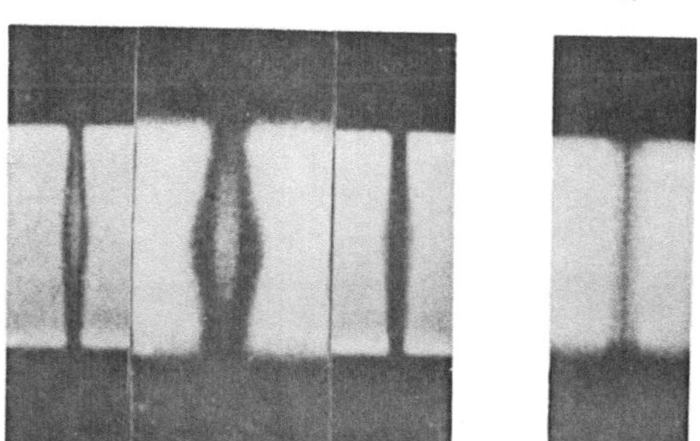

Fig. 68.

5', 6', 7'. Types of Magnetic Resolutions in Non-uniform Fields.
10'. Superposition of Magnetic Components.

polariscope, however, leaves it undecided whether the obliquity of the phenomenon is really due to the outer components.

104. The position of the plane of maximum polarisation can be determined fairly accurately. The obliquity of the major axis of the outer ellipses of sextet and quartet in one experiment was 5°; with the very same vapour-density and the same strength of field, the plane of partial polarisation made an angle of 21° with the vertical. At first sight it seems rather startling that the polariscope of Savart is so sensitive to the obliquity of the ellipses.

The phenomenon of the partial polarisation of the emitted light is very complicated and the complete theory is still outstanding. It does not seem doubtful, however, in what direction we have to look for the explanation of the remarkable difference between the indications of the two instruments; they measure different quantities.

As long as the inclination of the vibration ellipses of the emitted light is zero, the total light also vibrates symmetrically relatively to the vertical. If the inclination is not zero, however, but has the value α, the plane of maximum resultant luminous motion is inclined at an angle $\alpha+\beta$, which may be occasionally much greater. (See Zeeman and Winawer, *l.c.*, § 28.)

Oblique Position of the Vibrations of the Middle Components.

105. Whereas the inclination of the vibration-ellipses of the outer components could be demonstrated first for the sextet, it was for the quartet, on the contrary, that we first succeeded in verifying the second of Lorentz's above-mentioned conclusions (§ 100). The deviation of the vibrations of the middle components of the quartet from the horizontal line can be shown in the same manner as the inclination of the ellipses (§ 101).

The principal section of the nicol before the slit was placed at an angle of about 30° with the horizon. The outer components of the quartet of D_1 are then scarcely visible. The inner components are rather dark. The direction of the field may be indicated as direction 1. Under the influence of the reverse field 2, the middle components become more black. If the nicol be placed in the symmetrical position, then it is with the field-direction 1 that the middle components are most distinct. The angle θ in this experiment was 47°.

Two different attempts to measure the angle between the vibration and the horizontal gave the results 4°.5, and 5°.5. These measurements are very difficult, however, and perhaps indicate only the order of magnitude of the inclination. The vicinity of the outer components greatly interferes with the accuracy of the adjustment of the nicol, for while it is moved about near the position of extinction and approaches to a vertical direction, the greater intensity of the outer components distracts the eye.

106. We have made yet another experiment which confirms the result of section 105 for both the sodium lines and also exhibits the relation between the inclinations of the different components. This connection is shown diagrammatically for a triplet in Fig. 63, for the result obtained with the middle components of the quartet and the sextet can certainly be applied qualitatively to the triplet.

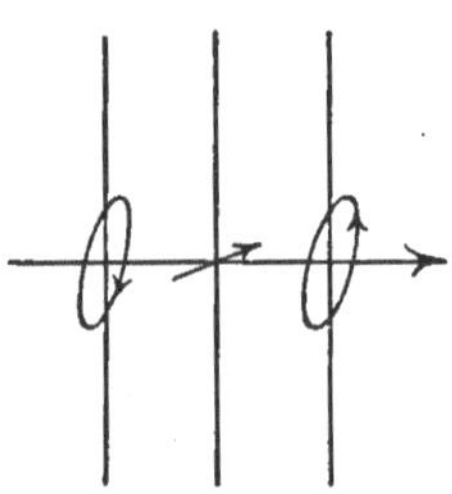

FIG. 63.

The experiment was as follows: the principal section of the nicol made an angle of +40° with the vertical; the positive direction in Fig. 63 was counter-clockwise. Then the nicol was placed at 320° (*i.e.*, in the symmetrical position). The last position may be indicated as position *B*, the first mentioned as position A. The direction of the field remains unchanged. In position A all lines were weaker than in position B.

Hence we conclude that the ellipses as well as the vibrations of the middle components are inclined; moreover, that the relative position of the vibrations must be that shown in Fig. 63.

107. In an important paper published some years ago, Righi[1] says that Voigt's theoretical investigation of the general case of propagation of light in a direction inclined to the lines of force was published too late to guide him in his investigation. Righi expresses the opinion that it is somewhat improbable that in the course of his numerous observations peculiarities in the behaviour of the middle components as indicated by Voigt could have escaped him, and that Lorentz's elementary theory is in accordance with all the observed phenomena.

This seems in contradiction with our experiments. This contradiction vanishes, however, if we assume that the vapour in Righi's experiments was very dilute, or the field so intense that the components were neatly separated. In such circumstances our observations also are in complete accordance with the elementary theory, at least as to the polarisation of the components and the direction of the vibrations. Neither was it in Righi's experiments a matter of course to reverse the direction of the magnetic field, the procedure which most easily exhibits any obliquity of the vibrations.

108. Another method of demonstrating the oblique position of the vibrations of the different components is the use of a half-wave-length plate. Vibrations from the source, making a definite angle with the principal direc-

[1] *A. Righi*, Sul fenomeno di *Zeeman* nel caso generale d'un raggio luminosa comunque inclinato sulla direzione della forza magnetica. *Mem. di Bologna*, 1899.

tions of the plate, after traversing it, are rotated through twice that angle. It is easily arranged, therefore, that the field of view is divided into two parts, the obliquity being shown by a difference of intensity of the magnetic components in the two halves of the field.

Connection between the Inclination of the Ellipses in Particular Cases.

109. The direction of the magnetic field, and that of propagation of the beam traversing the magnetised source of light, determine the sense of the inclination of the vibration-ellipses (§ 102). If the direction of the field be reversed, the sign of the inclination of the vibration-ellipses also changes In Fig. 62 (§ 102) the connection established by our experiments between the three mentioned directions is given.

Let OF be the magnetic force, and let the beam, traversing the magnetised flame O, be propagated in the direction from O to S. The inclination of the ellipses in this case is indicated in Fig. 64. The plane normal to the ray and containing the ellipse has been rotated round the dotted line until brought into coincidence with the plane of the paper.

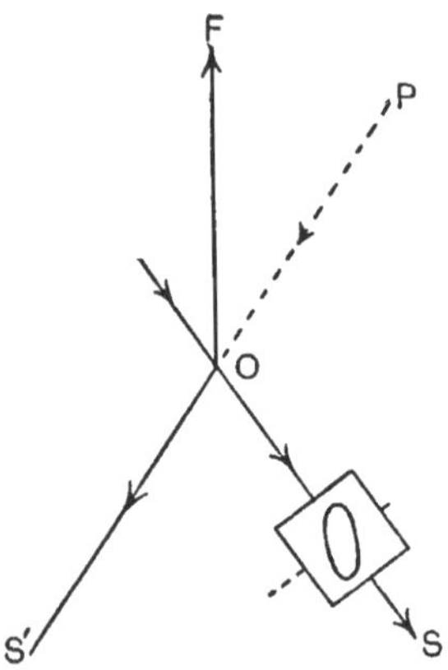

FIG. 64.

What is the inclination, if the source of light be traversed by the beam in the direction OS'?

This question is easily answered by applying the well-known method of reflected images. The geometrical outlines of all things composing a given system, together with the physical processes in the system, which we suppose may be all represented by geometrical figures, we imagine reflected *at every instant* in a plane V. The new system obtained by reflection, which we call the image of the original system, is a possible one, as soon as the last-mentioned one has an objective existence.

Applying this to our experiment and placing the plane V parallel to OF and perpendicular to the plane of the paper, we obtain from system I the system II (Fig. 65).

The magnetic field in the second system is the inverted image of the field in the first; indeed, before taking the

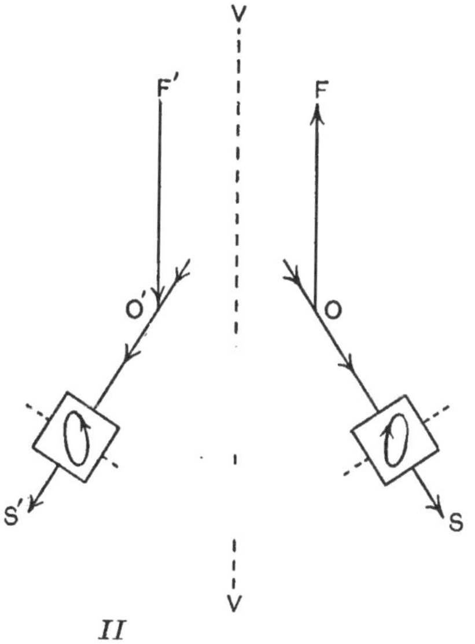

FIG 65.

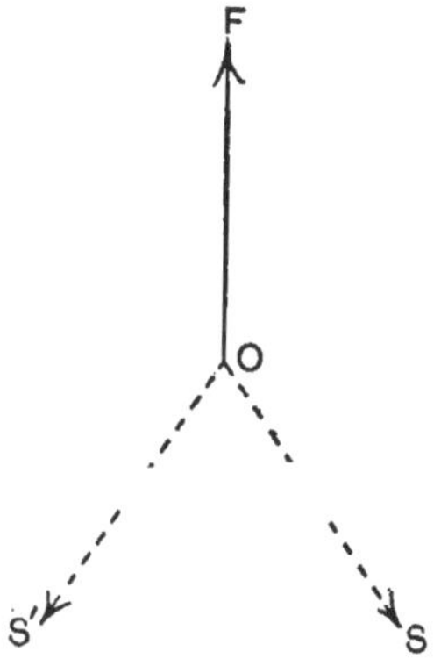

FIG. 66.

image of the field we have to substitute it by the equivalent ampere currents. Hence in II the arrow $F'O'$ is drawn from F' to O'. The field in system II being afterwards reversed, the inclination of the ellipse changes its sign. Hence we conclude that (Fig. 66) if OF be the direction of the magnetic field, the inclination of the major axes of the ellipses, as observed from S as well as from S', is always from the lower left to the upper right quadrant.

By means of Savart's polariscope all this could be experimentally verified. We come to the same conclusion by using the experimental result of § 103, concerning the inclination of the ellipses in the beam *emitted* in the direction OP (see Fig. 64).

The close connection existing between emission and absorption enables us to predict the phenomena to be seen if light traverses the source in the direction OS'.

The third theoretical prediction which admits of being experimentally verified is

The Existence of an Angle θ_1.

110. Some experiments relating to the chromium *emission* line 5206 proving the existence of an angle θ_1 were made by Elias.[1] For $\theta = 12°$ the middle line of the triplet was very faintly visible, the direction of vibration made an angle of about 30° with the horizontal. For $\theta = 10°$ a trace of the middle line could still be seen; the angle between the direction of the vibration and the horizontal line seemed to amount to about 40°. For $\theta = 17°$ the angle between the two mentioned directions amounted only to 15°. For greater values of θ the direction of vibration was nearly horizontal. We must conclude that in this example θ_1 was about 10°.

While seeking in the *inverse* effect for manifestations of the existence of an angle θ_1, Zeeman and Winawer discovered an interesting phenomenon first observed at $\theta = 16°$.

By means of a quarter-wave plate and a calcspar rhomb a double image was obtained, the oppositely polarised vibrations being separated. With dense sodium vapour, the phenomenon resembles the pure longitudinal one. No trace of the middle components is visible.

If the density of the vapour be increased, however, to the limit obtainable by the introduction of a glass rod charged with melted salt into the gas-oxygen flame, new

[1] Elias, *Proc.* Acad. Amsterdam, September, 1910, 391.

middle components appear in the parts of the field of view originally bright. These new components are *unpolarised.*

In the case of D_1 their position could be measured fairly accurately. They coincide with the interior components of the D_1 quartet. An analogous result would certainly be found with the new components of D_2, but the accuracy of the measurement is here necessarily less.

The observation at $\theta = 16°$ was a rather difficult matter. In the memoir cited the arrangement used is described and the appearance of the line D_1 illustrated by a drawing.

Since then the present writer has obtained photographs of the phenomenon, one of which is reproduced in Fig. 67 (Plate VIII). The latter investigation was facilitated by the use of the admirable oblique-vision pole-pieces designed for the purpose by du Bois.[1]

The result that the new components are unpolarised seems paradoxical, because one has now become accustomed to expect polarisation of all magnetically-separated and displaced lines. It may be remembered, however, that according to Lorentz, in the case of a triplet the two elliptically polarised principal beams, which are transmitted between 0° and θ_1, have equal indices of absorption. Hence we may expect that in the circumstances of the case a magnetised vapour can produce in a continuous, unpolarised spectrum only unpolarised absorption lines.

The existence of the unpolarised lines would thus be explained, if it be permitted to apply to the middle components of a quartet the theoretical inference drawn for the central component of a triplet.

In a recent letter to the author, Voigt pointed out some difficulties inherent in such an interpretation. The subject merits reconsideration, and I hope to return to it on a suitable occasion.

[1] *Proc. Acad. Amsterdam*, **13**, 389, 1910.

Application of Results to the Interpretation of Sun-spot Spectra.

111. The various types of spot lines are indicated in Mitchell's diagram published in the *Astrophysical Journal*, **22**, 6, 1905, and copied with some modifications in the *Transactions* of the International Solar Union. This last diagram is reproduced on Plate VII of the present work.

The widening of the Fraunhofer lines as they cut across the spot is strikingly analogous to the appearance of the absorption lines which I have obtained in the laboratory by placing a flame in a *non-uniform* field.

The remarkable resemblance between the absorption lines in the spectra of sun-spots and the separations in non-uniform laboratory fields may be seen by comparing Fig. 68 (Plate VIII) with the diagram Fig. 57 (Plate VII). The most characteristic line types of the diagram are 5, 6, 7, and 10.

In Fig. 68 are represented some corresponding separations observed in the laboratory *without* nicol or other analyser, 5′, 6′, 7′ have been taken in non-uniform fields. 5′ is the quartet of D_1 observed across the field; 6′ the sextet of D_2 observed axially in a non-uniform field, very strong in the central part; 7′ also refers to D_2 in a weaker field, the observation being made across the lines of force. The type 10′ refers to the D_2 line, when observed in a direction parallel to the field. The field is uniform. The separation gives an example of the superposition phenomenon mentioned in § 30.

The analogy of the type 10′, Fig. 68, and the type of the "winged line" seems very remarkable. Of course, observation of the state of polarisation would be necessary in order to prove the analogy, and it certainly does not exist in such cases as the D-lines or the H and K lines.

112. If the elementary theory were strictly applicable to the magnetically divided lines of the sun-spot spectrum, it would be possible to determine the magnitude and the direction in space of the magnetic force.

The magnitude of the force is directly proportional to the magnetic separation; so that, if the separation of the line under consideration has been measured in a laboratory field, the intensity of the magnetic force is known at once.

From the ratio of the intensities of the middle line of a normal triplet and those of its outer components we can deduce θ the angle between the line of vision and the direction of the magnetic field. Indeed, the intensity of the outer components is independent of the direction of observation, and the intensity of the middle component is proportional to $\sin^2 \theta$. For $\theta = 90°$ the intensity of the middle line of the triplet is twice that of one of the outer components.

The projection of the magnetic force on the plane of the solar disc is found immediately, for it is parallel to the short diagonal of the nicol when placed in such a position that the middle line of the triplet is of maximum darkness.

113. All this applies only if the components are neatly separated, which is not so in the case of sun-spot lines. In drawing maps of the magnetic fields in sun-spots, showing the intensity, the direction, and the polarity of the magnetic force, the determination of the direction of the force gives some difficulties. We must now apply the results of the general theory.

The direction of vibration of the middle component does not indicate now the direction of the projection of the magnetic force. The angle between the two directions may even attain 40°.

The rule which determines the direction of the deviation may be indicated here. The direction of rotation in the vibration-ellipses of the outer components towards the red and towards the violet shows whether θ is acute or obtuse. If θ is obtuse (Fig. 62), then the relative position of the directions of the magnetic force, of the major axis of the vibration-ellipses, and of the vibration of the middle component is shown in Fig. 63.

From any point O draw a line OB parallel to the major axis of the vibration-ellipses of the outer components and a line OM parallel to the vibration of the middle component, the angle BOM being always chosen acute. The projection OF of the magnetic force on a plane normal to the line of sight then makes a positive acute angle with OB, the angle BOF being greater than BOM, the positive direction being reckoned from OB to OM.

By ascertaining whether or not the major axes of the ellipses and the vibrations of the middle component are perpendicular to each other, we can make sure whether the elementary theory may be applied or not. As the axes of the ellipses of vibration are also inclined relatively to the field, we cannot infer the position of the latter from that of the former.

It is true that the obliquity of the ellipses in our experiments was only small, even when the vibrations of the middle component made a large angle with the horizontal. In order to determine the direction of the projection of the magnetic force on the solar disc, it would seem, therefore, the best course, theoretically at least, to ascertain the position of the major axes of the vibration ellipses of the outer components. The direction sought for is then nearly perpendicular to the major axis.

114. The results of the present chapter, although complicating somewhat the interpretation of the observed phenomena, probably afford a simple explanation of some phenomena provisionally attributed by Hale to the rotation of the plane of polarisation of the light emitted by the central line when passing outward through the spot vapours.

In some cases, when a nicol is used, the central line of a spot triplet is present on one side of the spot and absent on the other. Rotation of the nicol through 90° reverses the appearance. If we admit that at one side of the umbra the magnetic intensity and the width of the lines are such that θ_1 is greater on one side of the spot than on the other, the change of the direction of the vibration of the middle line is at once explained.

CHAPTER X

CHEMICAL ELEMENTS AND MAGNETIC RESOLUTION. CONTRIBUTIONS TO THE CONSTITUTION OF THE ATOM

115. THERE is, in our opinion, no fact that shows in a more striking way that the chemical elements have something in common than the magnetic resolution of spectrum lines. All the line-spectra of the numerous elements which have been investigated up to the present are modified by magnetic forces. Numbers of spectrum lines can be split up into two, three, or more polarised components, and thus bear witness to the presence of vibrating electrically charged (*i.e.*, negatively charged) electrons, or systems of electrons, in the atoms of the examined elements.

For elements rich in lines we may convince ourselves that the resolution into triplets occurs most frequently. The table on the following page, from an important paper by King,[1] is very instructive in this respect.

About five-eighths of the lines in the spectra of iron and titanium separate into triplets, and therefore render a computation of e/m at once possible. These lines, which *do not* (so far as we yet know) *belong to series*, give rather divergent values for the amount of the resolution, and hence also of e/m.

Cotton[2] has examined the resolutions which occur most

[1] King, Contributions Mount Wilson Solar Observatory, No. 56, 21.

[2] Cotton, Société franc. de Phys. Séance du 7 Mai, 1909.

frequently, in the results of Moore, Van Bilderbeck-Van Meurs, Jack, and others. It appeared that the amount of resolution which corresponds to the value of e/m for cathode rays—the "normal" resolution—is not the most prevalent one. The amount of the separations seems to group regularly round some special values, which are found again for different elements.

From her measurements Mrs. Van Bilderbeek-Van

SUMMARY OF TYPES OF SEPARATION.

Separation.	Iron.	Titanium.
Unaffected	9	4
Triple	393	291
Quadruple	49	28
Quintuple	7	5
Sextuple	118	77
Septuple	37	12
Octuple	6	11
9 components	9	3
10 components	7	7
11 components	2	2
12 components	4	2
13 components	2	0
Unclassified	19	16
Total	662	458

Meurs[1] has derived that, for iron, 1·16 and 1·5 times the normal resolution occurs most frequently. From the triplets measured by Moore in the spectra of yttrium and zirconium, and by Jack in tungsten and molybdenum, she finds, if a denotes the normal separation,

maxima for Y at 1·13a
,, ,, Zr ,, 1·14a and 1·50a
,, ,, W ,, 1·13a ,, 1·40a
,, ,, Mo ,, 1·11a

The grouping of the separations is shown still more

[1] Thesis for the Doctorate, Amsterdam, 1909.

clearly in a curve (Fig. 69) showing the frequencies of the occurrence of the different magnitudes of separations. The first curve refers to iron (138 triplets), the second to nickel (163 triplets) according to the observations of Miss Graftdÿk,[1] whose curve for nickel exhibits maxima at 1·10, 1·17, and 1·41 times the normal separation. In the examination of the curves it should be borne in mind

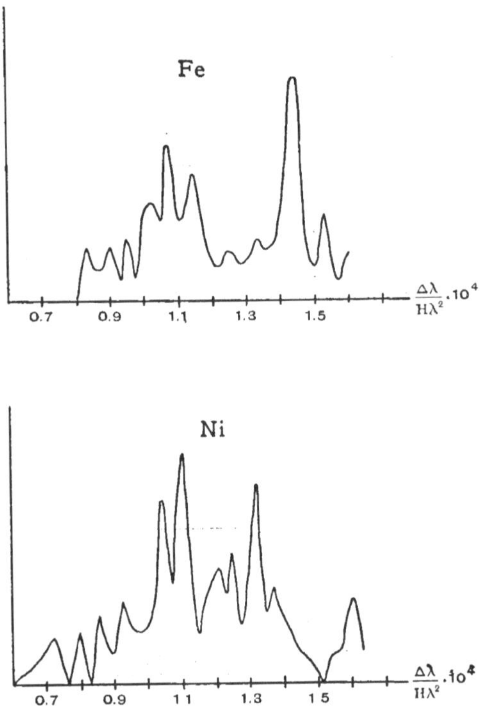

FIG. 69.—FREQUENCY CURVES OF MAGNETIC TRIPLETS.

that for a "normal" triplet $\Delta\lambda/H\lambda^2 = 0{\cdot}94 \times 10^{-4}$, if $\Delta\lambda$ represents the distance of the outer components, and the wave-lengths are expressed in centimetres.

Analogous results have been found by Babcock,[2] who partly used the observations of the above investigators. He took, however, King's observations for iron, his own

[1] Thesis for the Doctorate, Amsterdam, 1911.

[2] Babcock, *Astrophysical Journal*, **34**, November, 1911.

for vanadium and chromium, those of Miller for manganese, and those of Reese for nickel. For eight of the thirteen spectra investigated the curves ought to be shifted over an amount of 0·03 towards the right in his figure, because a correction has to be applied to the field-intensities used by Miller and Moore[1] (Fig. 70).

For entirely different elements we get, accordingly,

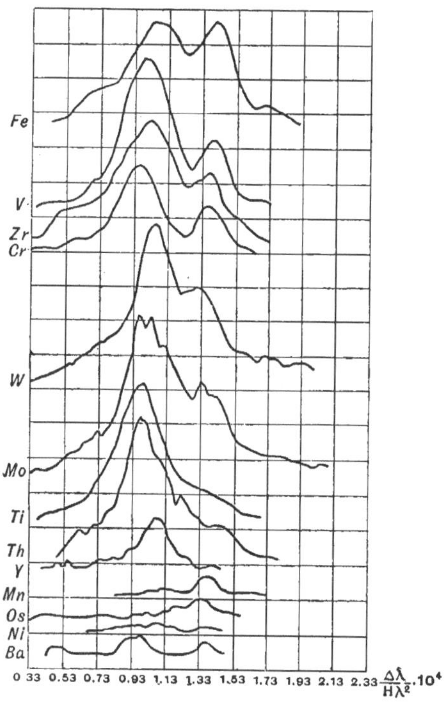

Fig. 70.—Frequency Curves of Magnetic Triplets.

values of e/m which never depart much from $1·8 \times 10^7$. The atomic weights of the elements and the specific charges of the ions diverge very much more (H = 1, U = 238).

It is easily possible that all the lines which do not belong to series will eventually appear to be emitted by identical vibrating electrons. There are various reasons

[1] Cf. Cotton, *Journ. de Physique*, Feb., 1912, p. 97; Van Bilderbeek-Van Meurs, *loc. cit.*, p. 65, and *Archives Néerl. Ser.*, (2), **15**, p. 395, (1911).

to be urged in favour of this; in consequence of connections between electrons or their interaction the apparent mass may possibly be now enlarged, now diminished.

Closer and sharper are the relations which manifest themselves for lines *that belong to the same series. Such lines exhibit identical magnetic resolutions not only for one and the same element, but also for elements which are chemically entirely different*, if the scale of frequencies be used (*cf.* § 40).

116. Before describing the results of the experimental investigation, we give the periodic table of the elements, to which we shall have occasion to refer repeatedly.

Period	Ser.	0	I	II	III	IV	V	VI	VII	VIII		
1st	1	—	—	—	—	—	—	—	H 1·008	—	—	—
	2	He 3·99	Li 6·94	Be 9·1	B 11·0	C 12·00	N 14·01	O 16·00	Fl 19·0	—	—	—
2nd	3	Ne 20·2	Na 23·0	Mg 24·32	Al 27·1	Si 28·3	P 31·04	S 32·07	Cl 35·46	—	—	—
3rd	4	Ar 39·88	K 39·10	Ca 40·09	Sc 44·1	Ti 48·1	V 51·2	Cr 52·0	Mn 54·93	Fe 55·85	Co 58·97	Ni 58·68
	5	—	Cu 63·57	Zn 65·37	Ga 69·9	Ge 72·5	As 74·96	Se 79·2	Br 79·92	—	—	—
4th	6	Kr 82·9	Rb 85·45	Sr 87·63	Y 89·0	Zr 90 6	Nb 93·5	Mo 96·0	—	Ru 101·7	Rh 102·9	Pd 106·7
	7	—	Ag 107·88	Cd 112·40	In 114·8	Sn 119·0	Sb 120·2	Te 127·5	I 126·92	—	—	—
5th	8	X 130·2	Cs 132·81	Ba 137·37	La 138·0	Ce 140·0	Nd 143·0	—	Sa 150·3	—	—	—
	9	—	—	—	—	—	—	—	—	—	—	—
	10	—	—	—	Yb 172·0	—	Ta 181·0	W 184·0	—	Os 190·9	Ir 193·1	Pt 195·2
	11	—	Au 197·2	Hg 200·0	Tl 204·0	Pb 207·1	Bi 208·0	—	—	—	—	—
7th	12	—	—	Ra 226·4	—	Th 232·4	—	Ur 238·5	—	—	—	—

It has long been known that relations exist between the line-spectra of the elements and their atomic weights. This is admirably expressed in a graphic representation of the spectra, borrowed from Kayser's "Handbuch," Vol. II. (Fig. 71, next page).

In a definite group of elements the spectra are displaced towards the side of the red with increasing atomic weight. The same aspect is obtained for all the elements; and after the formulæ for the series have been found it is even possible to point out for each of the lines in a spectrum corresponding lines in another. The analytic function, however, which expresses the limits of the subordinate series of the constant A of the formulæ of Kayser and Runge, or of Rydberg, in terms of the atomic weight, is not known. We only know that the relation between the series limits and the atomic weight is expressed by a wavy curve; the term periodic function, which is some-

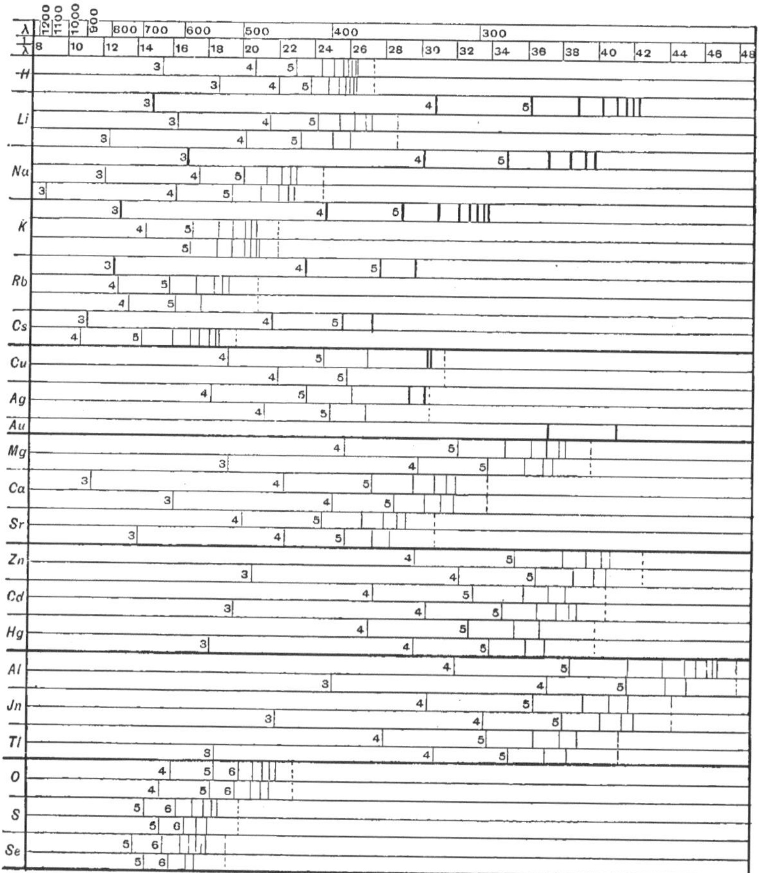

FIG. 71.

times used in this case, and in all other considerations on the periodic table, must not be understood here in the sharply defined sense which mathematicians attach to it. Reinganum has directed attention to a possible relation of the serial spectra to the atomic volume.[1]

[1] *Physik. Zeitschr.*, **5**, 302, 1904. *Cf.* also Rossi, *Phil. Mag.*, **22**, 922, 1911.

Scales for the frequency and for the wave-length are given at the top of the drawing. For doublets and natural triplets, only the first lines have been drawn so as not to crowd the figure. The principal series and the two subordinate series have each been drawn in a separate strip. For the alkalies there are three of these regions. For Cs the second subordinate series is not given, because it was found after Kayser's diagram had been made. Also the Bergmann series, which were discovered later, and lie almost entirely in the red and the ultra-red, are not represented.

In column I of the periodic table occur Cu, Ag, and Au. For the two first metals, series are known, but not for gold. All three have a very intense pair of lines which Rydberg takes as the first member of the principal series, which is represented in the figure by thick lines. The theoretical limits of the subordinate series are indicated by dotted lines. It has been proposed by Schuster[1] to call the point to which a series converges the "root" of the series. In these roots the displacement with the atomic weight is particularly marked.

Rydberg and Kayser and Runge have pointed out that the differences of the frequencies of lines which occur as pairs or natural triplets, increase with the increase of the atomic weight. These differences of frequency are for a group of chemically allied elements approximately proportional to the squares of the atomic weights.

117. We shall now give a survey of the principal types of magnetic resolutions for different elements. For this purpose graphical figures drawn on the scale of the frequencies, and at the same time indicating the normal resolution may be used. Such figures have been employed by Voigt,[2] and particularly by Cotton,[3] to represent the results of the measurements in a very concise form. In

[1] Schuster, "The Theory of Optics," Second Edition, 299, 1909.

[2] *Magneto-optik*, 87. [3] Février, *Le Radium*, **8**, 1911.

the two accompanying collective figures (Figs. 72 and 73) vibrations at right angles to the field are represented by full lines, and vibrations parallel to the field by broken lines. In the few cases in which the two kinds of vibrations coincide, one half of the line is drawn full and the other broken. The finely dotted lines serve to fix the scale; they are drawn at distances equal to half the normal separation, but they are never drawn in the resolution figures themselves.

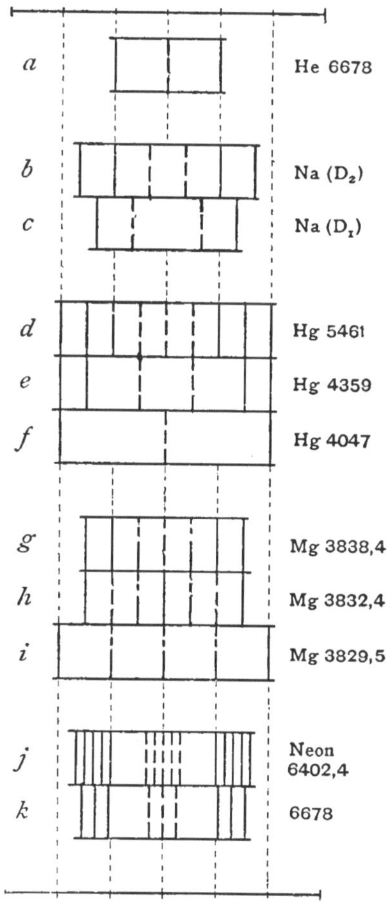

FIG. 72.—TYPES OF MAGNETIC RESOLUTIONS.

In the *transversal* effect, the *ensemble* of the lines is observed for a given spectrum line. In observations made in the direction of the field—the *longitudinal* effect—only the continuous lines remain. The figures summarise the principal characteristics of the resolution, when we remember also that the circular vibrations which have the direction of Ampère's currents in the electromagnet are situated on the violet side of the original line. In this case the sign of the effect is sometimes called *negative*. In many cases it had already been ascertained that for line spectra the sign is negative. In an investigation undertaken for this purpose in collaboration with Winawer for a number of complicated resolutions, I have always found this rule confirmed. The sign of the effect is positive only for a very few spectrum lines which are not to be attributed to atoms, as for the majority of gas-spectra, but to molecules (see below). The only thing these figures do not give is the ratio of the intensities and the aspect of the components. On the other hand, as Cotton remarks, they can take the place of the numerical data, for the magnetic fields used have but very rarely been determined with greater accuracy than to 1 per cent.

According to § 40 the distance of the outer components for a normal or standard triplet is expressed by the equation $\Delta\lambda = 0{\cdot}94 \times 10^{-4} \times H\lambda^2$. The resolution for a given field can now be found for each of the lines occurring in the figure. Thus we read, for instance, that the distance of the outer components of the quadruplet of D_1 amounts to four-thirds of the normal distance. Hence in a field of, *e.g.*, 20,000 gauss this distance becomes $\Delta\lambda = 4/3 \times 0{\cdot}94 \times 10^{-4} \times 2 \times 10^4 \times 5896^2 \times 10^{-16} = 0{\cdot}870$ Ångströms.

118. The examples *a—i, p, q, r*, refer to lines *which are connected in series*. As we saw in Chapter IV., the rule which has been proved by Preston and Runge and Paschen that they must exhibit resolutions which are *identical* and *of equal amount* when measured on the scale of frequencies, holds for such lines.

Example *a* refers to helium. We put this first, because, as Lohmann has found, the nine principal lines of the visible spectrum of this element are split up exactly according to Lorentz's elementary theory. The outer components of the helium triplets are at normal distance, corresponding to the ratio e/m of charge and mass of the electron of $1{\cdot}77 \times 10^7$. The nine lines referred to belong to five of the six series of helium which are known; all the series behave here in the same way.

Another example of an element which behaves in this simple way is probably hydrogen.[1] In this connection, it

[1] Croze, *cf.* § 121.

is remarkable that according to Voigt the red Li-line, which belongs to a principal series, is changed into a normal triplet. This, however, is not what might be expected; for in the case of the related alkali metals (Na, K, Rb, Cs), there are two principal series, one corresponding to D_1, the other to D_2. According to a rule mentioned above (§ 116) the distance of double lines is nearly proportional to the square of the atomic weight, and the red Li-line may be expected to be a double line consisting of two components at a distance of 0·6 Å.U. Nothing has been observed of this (see, however, p. 173), and there is therefore already a difference between the spectrum of lithium and that of the other alkali metals outside the magnetic field.

Helium and lithium are the first elements of the first period of the periodic table. It seems, however, not impossible that the lines of all the elements which belong to the first period exhibit normal triplets in the magnetic field. This would render the study of Be, B, C, N, O, Fl very interesting, as that of elements of which the amount of the magnetic spectrum effect has not been measured.[1] *b* and *c* indicate the resolution of the lines which belong to the two *principal series* of sodium, one corresponding to D_2, the other to D_1. Identically the same resolution is found for the other metals which occur in the second column of Mendeléeff's table, viz., for K, Cu, Rb, Ag, Cs, and Au. But for lines of elements which are chemically less closely related the same sextuplets and quadruplets can also be found in a magnetic field, and thus it can be proved that the same principal series occur. Runge and Paschen have proved that Mg 2795 – Ca 3934 (K) – Sr 4077 – Ba 4554 – Ra 3815 behave like D_2, while the characteristic quadruplet of D_1 occurs also for the metals of the second column, for Mg 2802 – Ca 3969 (H) – Sr 4215 – Ba 4934 – Ra 4682. Purvis has added to this Sb 3638 and some lines of Pb.

[1] Cotton, *Soc. française de physique*, March 1, 1912.

Before the magnetic resolution had been discovered, Rydberg found that the principal series is in close relation with the second subordinate series, which relation reveals itself among others in the fact that the lines of a pair of the second subordinate series correspond in inverse order with the lines of a pair of the principal series. The same relation was expected to exist in the resolutions in a magnetic field, and this is really the case.

Thus the resolutions of the D-lines have been found for a number of pairs of lines, but the line with greater wave-length has the type of D_2. Hence, such lines are considered to belong to the second subordinate series, although this cannot be proved in the usual way by formulæ, for lack of sufficient lines in the region of observation.

Similar to the line D_2 (5890) are divided:

Tl, 5351; Mg, 2936; Ca, 3737; Sr, 4305; Ba, 4900; Ra, 5814.

Like D_1 (5896):

Tl, 3776; Mg, 2928; Ca, 3706; Sr, 4162; Ba, 4525; Ra, 4533.

The simple relation existing between the resolutions for D_2 and D_1 and with the normal resolution has been mentioned already; and is, moreover, clearly to be seen from figures *b* and *c*.

According to Kayser and Runge's nomenclature, two *subordinate series* are found in the spectrum of mercury vapour; both are threefold. Strictly speaking, there are twice three series shifted in the scale of the frequencies by a constant amount.

Runge and Paschen have very fully examined the magnetic resolution, and have been able to verify closely the simple relations which we have found for the lines of the (threefold) second subordinate series. The lines of the second subordinate series have no satellites, and the separations are great.

The types of *d*, *e*, *f* are those of the Hg lines 5461, 4359, 4047. The last line gives a triplet with a resolution which is twice the normal one. For the other lines, the relation with the normal one is also obvious; thus the distance of the outer components for type *d* is divided into eight equal parts. The intensities of the nine components thus obtained are not equal, the outer components being much weaker than the others.

The type described is found in spectra of the elements of the third column of the periodic table, as the following table shows:

Mg, 5184 ; Zn, 4810 ; Sr, 4436 ; Cd, 5086 ; Hg, 5460, etc.
Mg, 5173 ; Zn, 4722 ; Sr, 4362 ; Cd, 4800 ; Hg, 4359, etc.
Mg, 5168 ; Zn, 4680 ; Sr, 4327 ; Cd, 4678 ; Hg, 4047, etc.

The absolute measurements of P. Weiss and A. Cotton[1] refer to the zinc lines given here. It is to the magnetic resolution of these lines that we owe the first correct value of *e/m*. Cotton and Weiss's measurements give the same value, which was afterwards found from the latest accurate measurements with cathode rays by Classen and Wolz.

The results for the triple *first subordinate* series of the elements of the third column are much less simple. Lines of corresponding series of chemically related elements are no longer split up in the same way. This does, indeed, remain the case for the lines of the same series. Preston's rule, however, is no longer applicable. According to Miller,[2] the simplest case is that of the magnesium lines. The distances of the components are in simple relation to the normal resolution. The polarisation of several components, however, is no longer complete, as is also represented in the figures (*g*, *h*, *i*) Mg 3838·4, 3832·4, 3829·5. For Mg 3832·5 it is remarkable that the middle component vibrates normally to the lines of force. For

[1] *Journ. de Phys.*, **6**, 429, 1907.
[2] *Ann. der Phys.*, **24**, 105, 1907.

the corresponding lines of Ca, Zn, Cd, Miller has observed identical pure triplets, in which, however, the distances of the components are no longer in simple relation to the normal resolution. *p*, *q*, *r* refer to the Ca, Zn, Cd lines of the first subordinate series. For Hg Runge and Paschen have already investigated the resolution of the first subordinate series. The magnetic changes are then complicated, which is possibly connected with the greater number of satellites by which the Hg lines of the first subordinate series are accompanied.

It is not improbable that series will also be found in the spectra of other elements than those examined up to the present time. We can easily convince ourselves of the fact that series have been discovered only in spectra that are poor in lines. The number of lines of the arc-spectrum as function of the atomic weight is shown by Exner and Haschek[1] in an interesting graphic representation. All the elements in the spectra of which series have been discovered lie in, or close to the minima of the curve. So for the other spectra the greater number of lines has therefore probably been confusing and has hampered the discovery of series. The opinion has been expressed that these series may be of a different character from those represented by the formulæ of Kayser and Runge, and of Rydberg. Very recently Van Lohuisen[2] has found that in an important class of elements, for which this was not supposed, the same type of formula continues to be valid. We refer to the elements Sn, Pb, As, Sb, and Bi. For Sn and Sb Van Lohuisen has given numerical formulæ for the series, and this will probably also apply to the three other elements; for as Kayser and Runge[3] found, lines with constant differences of

[1] Exner and Haschek, "Die Spektren der Elemente bei normalem Druck," 37.

[2] "Thesis for the Doctorate," Amsterdam, 1912.

[3] Kayser, *Handbuch* **2**, 574.

frequency occur for all five elements, a fact of which Van Lohuisen made very important use in his investigations.

Very little is known as yet about the spectra of the five elements under consideration in the magnetic field. Purvis has, indeed, investigated a number of lines, but not so many as could be desired. Not before a more complete investigation has been made shall we be able to decide how far the elements of the fifth and sixth columns may be compared in regard to their magneto-optic behaviour with those of the first four columns. According to Purvis, Cu 3274 – Pb 3740 – Pb 2873 – Sb 3638, give identical quadruplets. They would, therefore, all exhibit type D_1, because we know this about the Cu-line.

119. We will now describe a few more or less isolated facts about the spectra in which as yet *no series have been discovered.*

The opinion that the lines which do not belong to series (or more accurately, probably, for which the series have not been found) are all split up into triplets, has proved erroneous, as also has the assertion that if resolution of this category takes place, it is the normal separation. Figs. 69 and 70 prove the contrary. In *s* the curve for Fe is once more represented greatly reduced according to the observations of Mrs. Van Bilderbeck-Van Meurs and Miss Graftdÿk (see above, Fig. 70).

From Fig. 69 it may be seen that twice the normal separation occurs exceedingly rarely. A pure triplet which exhibits a great resolution is represented in *l*, and refers to the Fe line 3476·83, according to the observation of Mrs. Van Bilderbeck. The Fe line 3497·99 is another example. The greatest magnetic resolution that has ever been observed in a line spectrum is that of the tungsten line 4269.6, which has been studied by Jack[1] (see *n*). There are thirteen components, and the outer (intense) components are at more than four times the normal distance.

[1] *Ann. der Phys.*, **28**, 1038, 1909.

According to Jack, another W line of wave-length 3868.1 has seventeen components, and exhibits a strange asymmetry, as it has two components which vibrate at right angles to the direction of the field, more on the red side than on the violet side (*o*).

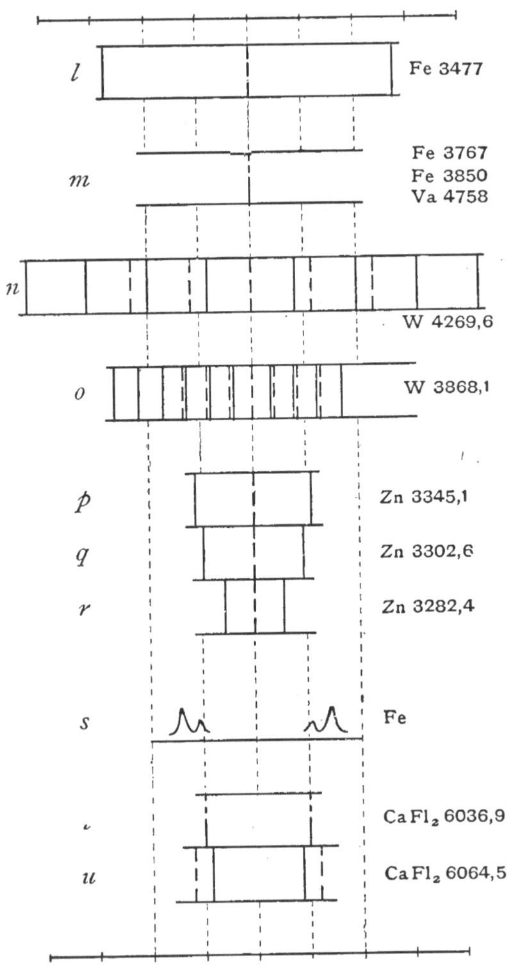

Fig. 73.—Types of Magnetic Resolutions.

In *j*, *k* two neon lines are represented. All the neon lines yield three groups of vibrations for the transversal effect, two vibrating perpendicular to, one parallel with, the field. Each of these groups may consist of one, two, three, four or five components. The distances of the components are in a rather simple relation to the normal distance.

As an illustration of the peculiarity that some lines are not affected at all by the field, *m* has been given. Under this head fall some iron-lines (H. Becquerel and Deslandres, Van Bilderbeck-Van Meurs). According to King's list there are nine such lines in the Fe-spectrum, four in the Ti-spectrum (see p. 157), and the list might be supplemented by some in other spectra.

The last figures, *t* and *u*, refer to band-spectra.

My first researches on the absorption band spectrum of iodine yielded a negative result. H. Becquerel and Deslandres could not find any change under the influence of the magnetic field either for the band spectrum of carbon (Swan spectrum) or for the cyanogen bands; nor did Purvis succeed in this for nitrogen bands. Dufour was the first to discover some band-spectra which are magnetically split up. These are spectra that are to be attributed to *molecules*. In *t* and *u* the bands for $CaFl_2$ 6064·5 and $CaFl_2$ 6036·9 are represented split up. To the last type belong some more bands of the fluorides and chlorides of the alkaline earths. A central component is found for none of these resolutions; there is, however, without doubt, a relation with the normal resolution.

In the case *u*, the circular polarisation is in the direction of the lines of force as usual, but for case *t* the effect is *positive*. In the latter case also, the complication presents itself that part of the light is always circularly polarised in the normal way, so that the vibrations do not seem circularly but elliptically polarised.

120. This is the place to mention a phenomenon which has been brought in connection with the magnetic resolution. Humphreys and Mohler[1] first proved that under the influence of external pressure many lines in the arc- and spark-spectrum are subjected to a slight displacement towards the side of the red, and are generally at the same time widened. In detail there is no close relation between resolution and pressure-shift, but there is no doubt a general relation between these two phenomena. Line-spectra exhibit both phenomena; band-spectra generally exhibit neither. After Dufour had found, however, that the bands of calcium fluoride were sensitive

[1] Humphreys and Mohler, *Astroph. Journ.*, **3**, 144, 1896; **4**, 175, 1896; particularly Humphreys, *ibid.*, **6**, 169, 1897.

to the magnetic field, Rossi[1] demonstrated that they were also subjected to a pressure-shift. Both effects differ for different elements and lines, and in general increase with increase of wave-length. Humphreys[2] has expressed the view that the pressure-shift owes its origin to a magnetic influence of an atom on the frequency of adjacent atoms. Since the luminous particle is affected by an outside field, it must also be affected by the magnetic fields of its neighbours. The intensity of the intra-atomic fields was estimated by Humphreys in 1906 at an order of magnitude of 10^8.

121. The foregoing survey will give at least some idea of what has been found about the magnetic resolution for different elements. For a number of elements our knowledge is still very deficient.

The results obtained so far may be summarised in a few sentences as follows:

In the first place, it appears clearly that there is an exceedingly close relation between the magnetic resolution and the properties of the electrons of the cathode rays. This relation is simplest in the case of the helium lines, but without any doubt it also exists in a number of other cases. Runge's rule points in this direction.

In the second place, the very important result has been obtained that there is identical resolution of all the lines of one and the same series for one element, and of the corresponding series for different elements; and in cases in which no series are known, identical resolution of isolated lines of different elements.

This conformity in numerical and sharply defined (polarisation) properties for lines belonging to the spectra of elements chemically so widely divergent, as *e.g.*, sodium, barium, thallium, lead and antimony raises a problem to which also the chemist can scarcely remain indifferent.

[1] Rossi, *Proc.* Royal Soc., **82**, 518, 1909.

[2] *Astrophys. Journ.*, **23**, 232, 1906; **35**, 268, 1912.

A few words may be said here concerning some very recent and valuable results due to Paschen and Back[1] (*cf.* § 86). These authors discovered that the lines of a very close series-triplet, or series-doublet, influence each other in a very peculiar manner. Under the action of a strong magnetic field we might expect to observe a superposition of the types of separation of the separate pair lines, but contrary to expectation only a triplet is seen. An example furnishes the pair line 2852·9 of sodium. The combined type of D_1 and D_2 was to be expected, but instead we get a triplet with nearly normal separation, the middle component vibrating parallel to the field, and the outer components partially polarised. Preston's rule here breaks down completely, for the (double) line 2852·9 belongs to the principal series of sodium.

Among the lines investigated are also the lithium lines. By analogy with the other alkali metals and their series many physicists regarded these lines as very close pairs, though others inclined to the opinion that the lithium lines were simple. While this volume is in the press, the author[2] has proved, however, that Li 6708 is a double line and it is now extremely probable that the other lines of the principal series of lithium are also double. Voigt obtained the result (§ 118) that the red lithium line (6708) is resolved into a triplet of at least nearly normal separation. Back's measurements relating to four lithium lines prove that all these lines give triplets which are normal within the limits of the errors of observation.[3]

[1] F. Paschen and E. Back. Normale und anomale Zeeman effekte. *Ann. d. Physik.*, **39**, 897, 1912.

[2] Zeeman, "The Red Lithium line." *Proc.* Acad. Amsterdam, Jan., 1913.

[3] In the paper mentioned, Paschen and Back give photographs of the magnetic behaviour of a natural triplet of oxygen and of that of the lines H_α and H_β of hydrogen. The theoretical aspect of the matter is discussed in the light of the theory of coupled electrons (*cf.* p. 187) by Voigt, *Ann. d. Physik.*, **40**, 368, 1913 (*cf.* Sommerfeld, *ibid.* 748).

122. In the case of the He-spectrum the negatively-charged electron which is drawn back to its state of equilibrium by a force proportional to its displacement from it, is sufficient at least for one line, for the explanation of the triplets and the other observed phenomena. If it is supposed that there exists only one single kind of electron, it is clear that all the He-lines must exhibit like separation. It is, however, difficult to see why the He-lines should be arranged in series, if a separate vibrating electron corresponds to every spectrum line.[1]

The same remarks as for He also apply to the spectra of other elements of which the lines are split up into normal triplets. Besides, to hydrogen[2] and possibly to the elements of the first period of Mendeléeff's table, this, according to Paschen,[3] probably refers also to those lines which belong to series of simple lines.

While in these simple cases one vibrating electron is assumed, the supposition naturally suggests itself to assume one type of system of electrons for different elements for the explanation of the repeated complicated resolutions. Then, as Runge[4] supposes, the different elements would distinguish themselves only by the positively charged particles and the forces they exert.

Even if it were possible to imagine a system giving a complicated resolution, the same difficulty as before would continue to exist with regard to the series. There seems no imperative reason for a serial relation. The serial relation is not the cause of the complicated resolutions, for there are connected series of triplets only. Complicated resolutions have only been observed for lines which occur in natural doublets and triple lines and combinations of such lines. Hence we may conclude with Paschen that

1 *Cf.* Lorentz, *Encycl. d. Math. Wiss.*, **5**, Part 3, 224.

2 Croze, *Compt. rend.*, **154**, 1410, 1912.

3 Paschen, *Ann. der Physik.*, **30**, 746, 1909; Royds, **30**, 1024, 1909.

4 Runge, in Kayser's *Handbuch*, **2**, 668.

the complicated resolutions are determined by the cause of the doublets,[1] etc., and not by the cause of the series.

There are accordingly three closely allied problems: Why do the elements give series spectra? Why are the spectrum lines split up into regular groups of polarised lines under the action of a magnetic field? and Why are the magnetic resolutions the same for lines of one series?

123. We cannot here give a summary of all the different electrical atoms which have been proposed in recent times. For the history of the electron up to 1901, the reader may consult W. Kaufmann's[2] exposition.

In its main features Lord Kelvin, in a publication in 1901, gave a description of an atom built up of electrical units. But neither spectrum series nor magnetic resolutions are discussed there, though Kelvin's interest in these problems never flagged.[3]

The recent advances in electrical science are evident from the atoms of Larmor,[4] J. H. Jeans,[5] J. J. Thomson,[6]

[1] Paschen, *Jahrbuch des Radio-aktivität*, **8**, 174.

[2] W. Kaufmann, *Phys. Zeitschr.*, **3**, 9, 1901.

[3] Extract from a letter dated May 22nd, 1906, from Lord Kelvin to the author: "Ever since I had the pleasure of meeting you and hearing your lecture in London at the end of March, I have been greatly exercised in trying to see a complete dynamical explanation of the quadrupling and sextupling of spectrum lines, which you have found, in rarer cases than the tripling as produced by magnetic force in the field where the light originates. Lorentz's dynamics is thoroughly satisfactory for the doubling and tripling, but something more is wanted for the quadrupling, etc. I am greatly pleased with Lorentz's fundamental hypothesis, which seems to me absolutely true, that the vibrations which directly originate waves of light are electrons or atoms of resinous electricity."

[4] J. Larmor, "Aether and Matter," 1900.

[5] J. H. Jeans, "The Mechanism of Radiation," *Phil. Mag.*, [6], **2**, 421, 1901.

[6] J. J. Thomson, "On the Structure of the Atom," *Phil. Mag.*, [6] **7**, 237, 1904; J. J. Thomson, "On the Vibrations of Atoms containing 4, 5, 6, 7 and 8 Corpuscles, and on the Effect of a Magnetic Field on such Vibrations" (1904); Cambridge Phil. Soc. *Proc.*, **13**, 39, 1906; see also G. A. Schott, "On the Electron Theory of Matter and on Radiation," *Phil. Mag.*, [6], **13**, 189, 1907.

H. Nagaoka,[1] Rayleigh,[2] Lenard,[3] Stark,[4] Ritz.[5] Our limits do not allow us to do more than touch upon some considerations concerning a few of the models, which represent the experimental facts concerning series and magnetic resolutions better than others.

124. Among proposed molecular model theories, the theory of Ritz has found many supporters. In this theory magnetic atom fields are required of the order of 10^8 gauss. Of the same order of magnitude are the field intensities required by Humphreys to account for the pressure-shift, and demanded by Weiss for the explanation of the magnetic properties of matter. While three different groups of phenomena lead to the supposition of intense atom fields, we are led by the study of the magnetic bodies and of the spectra of the elements to the introduction of molecular magnets, magnetons. It is known how Weiss, through his magnetic studies, has arrived at the hypothesis of a constituent common to a great number of magnetic atoms. It may not be impossible that Weiss's magneton is identical with the molecular magnet that Ritz assumed for the explanation of the law of Balmer and allied laws. Also, Lenard[6] is led to the assumption of linear elements in the atoms by his study of the emission of light in phosphorescence.

A rapid summary may now be given of the theory of Ritz on the origin of the serial spectra and of the

[1] H. Nagaoka, "Kinetics of a System of Particles Illustrating the Line and Band Spectrum and the Phenomena of Radioactivity," *Phil. Mag.* [6], **7**, 445, 1904. This paper contains the theory of the "Saturnian" atom.

[2] Rayleigh, "On Electrical Vibration and the Constitution of the Atom," *Phil Mag.*, [6] **11**, 117, 1906.

[3] Ph. Lenard, "Über Lichtemission und deren Erregung," *Ann. der Physik*, **31**, 641, 1910. Über Äther und Materie. zweite Auflage. Heidelberg, 1911.

[4] J. Stark, "Principien der Atomdynamik," I Teil. Die elektrischen Quanten (1910), II Teil. "Die elementare Strahlung," 1911.

[5] W. Ritz, "Magnetische Atomfelder und Serien spectren," *Ann. der Physik*, **25**, 660, 1908; Oeuvres, 98.

[6] *Cf.* p. 677 of the publication cited first in § 123.

magnetic resolution.[1] The laws of the spectrum series always refer to $1/\lambda$, that is, the frequency. Lord Rayleigh[2] has emphasised the fact that the laws of the electric and elastic vibrations always lead in the first instance to an equation with the square, and not with the first power of the frequency. This is due to the fact that the time occurs in the form sin νt, in consequence of which a factor ν^2 appears when the acceleration in the equations of motion is determined, *i.e.*, by a twofold differentiation with respect to t. Now Ritz has observed that, as the vibrations with magnetic forces do not depend on the place but on the velocity, the equations of motion will be of the first order with respect to the velocities.

When intra-atomic fields are assumed, it is easy to prove that it is possible to obtain vibrations of an electron with the frequency of the light vibrations. Thus, when an electron with the charge e and mass m is compelled to remain in a given plane, it describes there circular vibrations with the frequency eH/mc. In this, H is the component of the magnetic force at right angles to the plane of vibration, and c the velocity of light. For values of H of the order 10^8 the frequency becomes of the order of that of the light vibrations.

Ritz assumes molecular magnets and the electron moving in a plane which is rigidly connected to the corresponding magnet and perpendicular to the line joining the poles. The electron is then subjected to the difference of the field intensities which each of the poles gives at the place of the electron.

Let μ be the strength of the pole, l the length of the elementary magnet. Then, if the charge e is at a distance r from the adjacent pole,

$$H=\mu\left(\frac{1}{r^2}-\frac{1}{(r+l)^2}\right)$$

---

[1] An early initiative attempt for explaining the two classes of phenomena is due to E. Riecke, *Ann der Phys.*, **1**, 399, 1900.

[2] *Phil. Mag.*, [6], **11**, 123, 1906.

hence the frequency

$$\nu=\frac{\mu e}{mc}\left[\frac{1}{r^2}-\frac{1}{(r+l)^2}\right] \quad \ldots \ldots \quad (1)$$

The analogy with Balmer's formula

$$\nu=N\left(\frac{1}{2^2}-\frac{1}{n^2}\right) \; n=3,\ 4,\ 5 \quad \ldots \ldots \quad (2)$$

and Rydberg's formula

$$\nu=N\left(\frac{1}{a^2}-\frac{1}{(n+C)^2}\right) \quad \ldots \ldots \quad (3)$$

is at once evident.

If we put $l=ns$, $r=as$, then follows from (1):

$$\nu=\frac{\mu e}{s^2mc}\left[\frac{1}{a^2}-\frac{1}{(a+n)^2}\right] \; n=1,\ 2,\ 3 \quad \ldots \quad (4)$$

For $a=2$ this becomes identical with (2).

It follows from (4) that for $n=\infty$ the frequencies approach a fixed limit; also that the coefficient of $1/n^2$ is universal, as Rydberg has stated, if e/mc and the elementary magnets are assumed to be the same for all elements.

Accordingly, the lines of hydrogen are obtained as follows. Let a charge e be placed at a distance $2s$ from an elementary magnet of length s. If this charge is set vibrating in a suitable way, it yields the hydrogen line Hα. When a second elementary magnet is adjusted at the end of the first, we get Hβ, a third gives Hγ, etc.

The essential part of Ritz's explanation is that the vibrations take place in magnetic fields originating from two poles which can occupy different equidistant places[1] lying on a straight line. The elementary magnets and the electrons must be the same for all substances.

The different modifications of systems of 1, 2, 3 . . . . magnets must then either be found in the *different* atoms

[1] From the identification of Weiss's magneton with Ritz's elementary magnet, I derive for its length 10^{-11} cm. This is at least not in contradiction to the estimations for the diameter of a molecule 10^{-8} cm., and the diameter of an electron 10^{-13} cm.

of radiating hydrogen, or all these modifications may be imagined in one atom or as succeeding each other in time.

125. To account for the magnetic resolutions, Ritz assumes that, under the influence of an external magnetic field, the system will describe a kind of precessional movement round the lines of force as axis. Let H_0—the intensity of the internal field—be much greater than the external field H, and θ the variable angle between H_0 and H, and ψ the angle of a plane through the directions H and H_0 with a fixed plane.

Ritz assumes that the resultant motion of H_0 may be represented by:

$$a \begin{cases} \cos\theta = a_0 + a_1 \cos\omega t + b_1 \sin\omega t + a_2 \cos 2\,\omega t + b_2 \sin 2\,\omega t + \ldots \\ \psi = \psi_0 + \omega' t + a_1 \cos\omega t + \beta_1 \sin\omega t + \ldots \end{cases}$$

ω and ω' are assumed proportional to the external field, H; they measure the velocity of the oscillations of H_0 with respect to H, and of the rotation round the external force.

Calculation shows that if ν_0 is the frequency of the original line, the following vibrations arise under the influence of H:

$$b \begin{cases} \parallel H & \nu_0 \pm m\omega & (m=1, 2, 3, \ldots) \\ \perp H & \nu_0 \pm n\omega \pm \omega' & (n=1, 2, 3, \ldots) \end{cases}$$

All the frequencies of the resolved lines can be calculated linearly and with simple whole coefficients from *two* values ω and ω'.

As ω and ω' represent the frequencies which occur in the expansion into series according to the harmonic analysis which we have used, we may also say, according to Ritz, that the magnetic resolution of the spectrum lines for rotatory movements of the atom acts the part of an harmonic analyser. It is easy to see that the formulæ (b) represent many of the observed resolutions

N 2

perfectly. We get, *e.g.*, the case of the quadruplet by putting in (b) $m=1, n=0$.

Very complicated resolutions are also represented in a simple way. The line 6402.4 of neon (fifteen components) is obtained with the values

$$m=0,\ 1,\ 2. \qquad\qquad n=0,\ 1,\ 2.$$

The line ω 4269.6 (thirteen components) with

$$m=0,\ 1,\ 2. \qquad\qquad n=1,\ 2.$$

The two motions of rotation and of oscillation, which are supposed to take place in the atom, are periodic in themselves. The resultant motion will also be so when ω and ω' are in a simple rational ratio to each other. Instead of the formulæ (b) we now get:

$$\begin{array}{ll} \parallel H & \nu_0 \pm m\omega' \\ \perp H & \nu_0 \pm \omega' \pm n\omega' \end{array}$$

All the distances between the components are then expressed by one constant. We then get the rule of Runge, when the constant ω' is in a simple relation to eH/mc.

A pure triplet is obtained by supposing $\theta=90°$, $\psi=\psi_o+\omega' t$; the axis of the system of elementary magnets then describes with uniform velocity a cone about the direction H, H being perpendicular to H_o. The components of the resolved line have now the frequencies ν_o+eH/mc, $\nu_o, \nu_o - eH/mc$, hence the separation comes out twice that of a normal triplet.

So far as the magnetic separation is concerned, Ritz's theory has constituted the subject of two interesting articles by Cotton.[1] Cotton gives an excellent exposition of Ritz's theory, and adds some new considerations of his own, among others concerning the positive effect, the case

[1] "La théorie de Ritz du phénomène de Zeeman."—*le Radium*, **8**, 363, 1911. "L'effet Zeeman positif dans les gaz et la théorie de Ritz."—*le Radium*, **8**, 449, 1911.

that the circular polarisation is opposite to that generally observed.

126. Voigt[1] has critically discussed Ritz's theory. Voigt's objections are of two kinds. In the first place, the origin of the precessional movement, which Ritz requires, is difficult to understand. If it is caused or sustained by the motion of the free electrons in the source of light, it is to be expected that the temperature will also exert influence on the amount of the magnetic separation, of which nothing has ever appeared.

When, however, the precessional movement has once been accepted, the *results* of Ritz's theory present difficulties. Thus the separation must yield many components (§ 125), which decrease in intensity with increasing m and n. Yet there has never been found a trace of these higher components on the photographic plates, even with long exposures. To be able, however, to form a many-sided and unprejudiced opinion of the theory, Voigt has treated the inverse effect starting from Ritz's hypothesis. Instead of studying the *emission* of a single molecule as Ritz did, Voigt calculated the *absorption* in a magnetic field for a system of molecules. As experience shows that absorption and emission always correspond, the emission of a molecular system can be found. It now appears from Voigt's calculations that the precessional movement, and in particular the component represented by $\delta\theta/\delta t$, will not exert any influence on the absorption in the magnetic field. For different molecules the said term will simultaneously assume opposite values; hence in the result its influence will vanish by interference. Voigt's conclusion is that the hypothesis of Ritz cannot explain the complicated resolutions for which it has been devised.

It cannot be denied that these objections are weighty. They do not affect the explanation, however, which Ritz

[1] Voigt, "Zur Theorie der komplizierten Zeeman Effekte," *Ann. der Phys.*, [4], **36**, 873, 1912.

gives of the laws of the series. Though there is something artificial about this explanation, it is the best we have at the present moment.

There is another method of development of Ritz's idea, which is not without advantage. The angular velocity of precession ω' is found equal to the usual value $eH/2cm$, if the laws of rigid dynamics for the motion of a top are applied to the Ritz system of elementary magnets.[1] Ritz developed his theory on other lines, but the top analogy gives the normal magnetic effect. In this way we therefore get the result of the elementary Lorentz theory of the magnetic effect and, moreover, some insight into a mechanism explaining the serial laws. Voigt's objections make it clear that the difficulties encountering an extension of Ritz's theory to *complicated* resolutions are at once formidable.

127. In 1897, Lorentz[2] extended the theory which explains the doublets and triplets. In this extension the radiating particles are conceived as systems of unknown structure, over which electric charges are distributed in some way or other. Besides the electric charges are supposed to be firmly attached to the parts of the system. During vibrations of the system the charges will therefore be actuated by forces from the side of an external magnetic field.

Let the configuration of the system be determined by a certain number of general co-ordinates of Lagrange $p_1 \, p_2 \, . \, . \, . \, . \, p_n$. With suitable choice of the coordinates the equations of motion of a system as here considered then assume the following form :—

$$m_1 \ddot{p}_1 = -f_1 p_1 + c_{12}\dot{p}_2 + c_{13}\dot{p}_3 + \; . \; . \; . \; . \; . \; . \; + c_{1n}\dot{p}_n$$
$$m_2 \ddot{p}_2 = -f_2 p_2 + c_{21}\dot{p}_1 + c_{23}\dot{p}_3 + \; . \; . \; . \; . \; . \; . \; + c_{2n}\dot{p}_n,$$
etc.

---

1 G. A. Schott, "Electromagnetic Radiation." Cambridge University Press, 1912. K. Körner, *Ber. deutschen. physik. Ges.*, **15**, 69, 1913.

2 Lorentz, *Ann. der Phys.*, **63**, 278, 1897.

The dots above the letters denote single, the double dots twofold differentiation with respect to time. The constants $m_1, m_2 \ldots . m_n$ are mass coefficients; f_1, f_2, f_3 refer to the quasi-elastic forces, and the terms with $\dot{p}_1, \dot{p}_2$ are due to the magnetic field.

The coordinates p_1, p_2 and the corresponding degrees of freedom are independent of each other in the absence of the magnetic field. But as is expressed by the terms with $\dot{p}_2$, etc., there is coupling between the coordinates when the field is present.

The coefficients c are proportional to the strength of the field, and the following relations exist between them :—

$$c_{21} = -c_{12} \quad , \quad c_{32} = -c_{23} \quad , \quad \text{etc.}$$

These relations are an immediate consequence of the fact that the force exerted by a magnetic field on a moving charge is always at right angles to the line of motion.

From these equations Lorentz shows that if a spectrum line is resolved into a components, a of the fundamental vibrations must have the same frequency. A line which is resolved into a components is therefore already an a-fold line before the field exists.

Only a spectrum line which arises by three modes of motion of which the frequencies are the same can be split up into a triplet by a magnetic field.

As the observations show, the components of a resolved spectrum line are just as sharp as the original line, at least of the same order (see *e.g.*, Fig. 19); it might be interesting to institute a separate investigation on this point.

To explain the result, which is certainly approximatively correct, all the particles of a light source must be assumed to be influenced in the same way. Lorentz therefore assumes that the influence of the magnetic field on the frequency of vibration does not depend on the orientation of the atoms with respect to the magnetic force. In this respect the atoms may then be called isotropic. An inequality of the properties in different directions would give rise to a diffuse broadening of the line. The particles of the elementary theory satisfy the condition of isotropy.

128. This is also the case with J. J. Thomson's model of an atom. According to his assumption, an atom consists of a positive charge uniformly distributed over a spherical space, in which a certain number of negative electrons, *e*, are enclosed. Such configurations of electrons can be in stable equilibrium. The simplest instance of J. J. Thomson's atom is the case of four equal electrons, which are in equilibrium at the angles of a regular tetrahedron. The vibrations of this system in a magnetic field have been investigated by J. J. Thomson[1] and Lorentz.[2]

Following this line of thought, at first there seemed to be a possibility that an interpretation might be found of quartets, quintets, etc. This, however, is not the case. It appears that the system under consideration can only give rise to triplets and doublets with the same states of polarisation which are derived from the elementary theory.

Thomson considers the density inside the positive sphere to be constant; Lorentz, however, assumes the density to be a function of the distance to the centre of the sphere. Lorentz shows that this supposition opens the way to two remarkable possibilities. First, the resolution of the components of the triplets can be different from that required by the elementary theory. We get the real value of the magnetic resolution when we multiply by a certain factor the resolution which we should calculate from the same e/m according to the elementary theory. This factor depends on the way in which the density in the sphere varies with the radius. Another interesting result is that the factor can also be negative. This means that in the radiation along the lines of force the circular polarisation of the doublet

[1] See § 123. [2] "Theory of Electrons," 120, 292.

components becomes such as would be assigned to movable *positive* electrons by the elementary theory. This possible inversion of the sign of the effect, merely by a special configuration of negative electrons, seems of particular importance. The positive magnetic effect was observed by Jean Becquerel for some absorption lines of crystals which contain compounds of the rare earths, and by Dufour for some band-spectra. The explanation of these phenomena has been attempted in different ways, and that under review is very simple. Now, however, it may be remarked that the positive effect may also be expected for *atom* spectra. Yet to bring about the positive effect either a crystalline structure or a *molecule* spectrum seems to be required. This would be in favour of an interpretation which assumes different mechanisms for the positive and the negative effect.

Thus for an explanation of the positive effect Voigt has made the assumption that in some molecular systems under the action of an external magnetic field, electric motions originate in the inside of the particles of such a nature that an internal field opposite to the external one is formed.

129. For the interpretation of the complicated separations, Lorentz's general system of equations and the supposition of magnetically isotropic atoms has proved to be rather too general. Voigt has abandoned the supposition of magnetically isotropic, arbitrarily orientated particles, and modified Lorentz's system of equations. He supposes that the radiating particles are orientated under the action of the field. Instead of Lorentz's co-ordinates p (p. 182), Voigt introduces the components ξ, η, ζ, of the displacements according to three vertical axes. In one group of equations for the q electrons in an atom there only occur the displacements ζ parallel to the direction of the field; in a second group only the ξ, η, displacements. There are now magnetic connections between the different displacements in the direction of the lines of force, and between the movements in the plane of X, Y.

If there are neither external forces nor resistances the equations become

$$\begin{cases} m\ddot{\zeta}_1 = -f\zeta_1 + c_{12}\dot{\zeta}_2 \; \ldots\ldots\ldots \; + c_{1q}\dot{\zeta}_q \\ m\ddot{\zeta}_2 = -f\zeta_2 + c_{21}\dot{\zeta}_1 \; \ldots\ldots\ldots \; + c_{2q}\dot{\zeta}_q \end{cases}$$

with the conditions $c_{jk} = -c_{kj}$.
and for the ξ, η vibrations:

$$\begin{aligned} m\ddot{\xi}_1 &= -f\xi_1 + c'_{12}\dot{\xi}_2 + \ldots + c'_{1q}\dot{\xi}_q + h\dot{\eta}_1 + h_{12}\dot{\eta}_2 + \ldots + h_{1q}\dot{\eta}_q \\ m\ddot{\eta}_1 &= -f\eta_1 + c'_{12}\dot{\eta}_2 + \ldots + c'_{1q}\dot{\eta}_q - h\dot{\xi}_1 - h_{21}\dot{\xi}_2 - \ldots - h_{q1}\dot{\xi}_q \\ m\ddot{\xi}_2 &= -f\xi_2 + c'_{21}\dot{\xi}_1 + \ldots + c'_{2q}\dot{\xi}_q + h\dot{\eta}_2 + h_{21}\dot{\eta}_1 + \ldots + h_{2q}\dot{\eta}_q \\ m\ddot{\eta}_2 &= -f\eta_2 + c'_{21}\dot{\eta}_1 + \ldots + c'_{2q}\dot{\eta}_q - h\dot{\xi}_2 - h_{12}\dot{\xi}_1 + \ldots - h_{q2}\dot{\xi}_q \end{aligned}$$

with $c'_{jk} = -c'_{kj}$, $h_{jk} = h_{kj}$
h determines the action of the field on the uncoupled electron.

The coefficients c_{jk}, c'_{jk}, h_{jk} must be taken proportional to the magnetic force.

The above set of equations, to which terms expressing a resistance, and others for external forces, may be added, forms the foundation for Voigt's theory of the propagation of light in a system of atoms.

The terms indicating the influence of the magnetic field are such that they satisfy the general conditions which follow from the principles of energy and the considerations of symmetry. With a suitable choice of the constants, Voigt's equations are competent to explain all the resolutions observed. Moreover, they combine in a rigorous way the magnetic resolution with all the phenomena of linear and circular double refraction that are in connection with it, as was discussed in Chapter V.

An objection to the above system of equations is their want of lucidity. This is not to be denied; though it cannot be a conclusive objection. Many thermodynamically derived relations cannot yet be kinetically explained. In simple cases it is, indeed, possible, and here also an image may be formed of the action in a few cases.

The connections expressed by h are a consequence of the fundamental law which governs the motion of an electric particle in a magnetic field.

130. Lorentz[1] has given a simple example of a case in which the connections c are also accounted for. Of four equal electrons A, B, C, D, A and B can move only along straight lines, which are in the XZ plane and the YZ plane respectively, and run parallel to the Z axis. The electrons C, D can only move in the XY plane. The connections consist in this, that C and D are always on opposite sides at equal distance from O, and that $AC = BD = L$ remains constant.

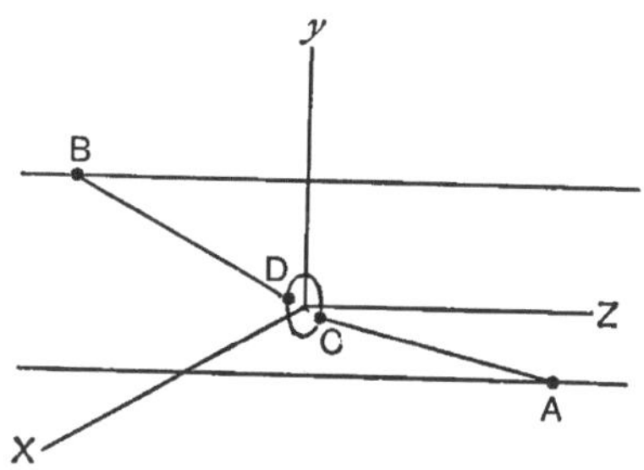

FIG. 74.

The displacements of the electrons are supposed to be infinitesimally small. The resultant motion gives a magnetic doublet with vibrations parallel to the direction of the Z-axis, *i.e.* of the magnetic force. The electrons C and D do not contribute to the resultant luminous motion, but they are influenced by the magnetic field.

131. Lorentz[2] has given another interesting instance which would account for the equality of the triplets for all the lines of a series.

Let a system of negative electrons in a positive sphere, as in J. J. Thomson's atom, be subjected to magnetic forces. If the positive sphere is immovable, the system of electrons will, as Lorentz demonstrates, if it was originally at rest, obtain a velocity of rotation :

$$k = -\frac{e}{2mc}H$$

[1] "Encyclopädie d. Math. Wiss," **5**, Heft 2, 220.

[2] Lorentz, "Encyclopädie d. Math. Wiss," **5**, Heft 2, 222; "Theory of Electrons," 123.

Now the electrons in the rotating system can execute vibrations. An important theorem of Larmor states that the same equations which hold for axes rotating with the system, also apply in the absence of the field with respect to stationary axes. In this way a magnetic resolution is obtained which agrees perfectly with that derived from the elementary theory. If the electric motions in the atom give rise to *a series* of spectrum lines, all these lines must be split up in the same way, because the velocity of rotation is the same for all the electrons.

An objection to this view, which was, indeed, already mentioned by Lorentz, is that systems which do not originate until the magnetic field is present would not get a rotation.

An objection of another nature may be derived from the laws of spectrum series. Thus, if Rydberg's formulæ are adopted, and if it is assumed that of the first subordinate series I, the part $1/(1+\mu_1)^2$ is changed by the magnetic field, and $1/(m+\delta)^2$ remains unchanged, the equal separation for all the lines of the series is explained. As the same expression $1/(1+\mu_1)^2$ occurs in the formulæ for the second subordinate series I, the separations of the first and the second subordinate series would have to be the same, which is not the case. However, against this reasoning it may be advanced that complicated resolutions always exist when the lines outside the field occur as natural doublets, so that a theory which gives triplets is not then applicable. For connected simple series of triplets, Lorentz's theory of rotating particles would hold good

132. The researches discussed in this volume refer to radiation in the magnetic field, but the results obtained give us information about the mechanism of emission and absorption for the general theory of light.

We cannot now doubt that visible and ultra-violet light is caused by vibrations of negatively charged systems of electrons of often very regular structure. In emission and absorption, in the phenomena of dispersion and propagation of these kinds of light, the same electrons are essential which have been discovered in the cathode rays, the electric current, and the β-rays of radium. It is very remarkable that the magnetic analysis of light points to a great stability of these systems of electrons. *The peculiarities of magnetic resolution remain the same whether,* e.g., *sodium vapour is made luminous in a spark, in a flame, or in a Geissler tube.*

133. We owe to Lenard and Stark very important researches on the way in which the emission of light is excited and maintained.

The disturbance brought about by the return of one or more electrons in the molecular or atomic structure would give rise to emission. At least this is the case in the phosphorescence of the alkaline earth phosphori.[1] For flames with metallic vapours this view presents difficulties. For them it is more plausible to assume, as Lenard[2] recently demonstrated, that the emission of light takes place at the impact (Nähewirkung) of free metal atoms with the molecules of the flame gases. If it is assumed that during this process temporary removal and rapid return of an electron of the metal atom takes place, the return of electrons would in this case also cause the excitement of the light emission.

Stark's discovery of the Doppler effect for canal rays seemed destined to indicate accurately how many electrons the radiating atoms in a vacuum tube have lost or gained. In the case of the series spectra of hydrogen and other gases the shift of spectrum lines pointed to a positive charge of the atom; in other cases no displacement could be observed.

[1] P. Lenard and Klatt, *Ann. der Phys.*, **15**, 671, 1904; Lenard and S. Saeland, *ibid.*, **28**, 476, 1909; Lenard, *ibid.*, **31**, 641, 1910.

[2] Lenard, "Sitzberichte d. Heidelberger Ak., 1911," Abh. 34, and the earlier literature mentioned there.

The further investigation of the "positive rays" by W. Wien, J. J. Thomson, and others, has shown, however, that there exists in those rays an "equilibrium" between charged and neutral particles. The uncharged luminous particles may attain a great velocity, which they acquired when they were temporarily charged.

An experiment by Von Dechend and Hammer[1] indicates clearly that in the same circumstances (so far as collisions are concerned), under which the neutral particles do emit light, the positive atom ions do not send out light. As Stark[2] has pointed out, we cannot draw any conclusion as to the charge of the atom that emits the spectrum.

The positive charges discussed in this section are the charges of the whole atom; the really *vibrating* part always continues to be negatively charged.

134. We will conclude this chapter, which treats of the magnetic resolution in connection with the *chemical* nature of the elements, with a consideration of a chemical nature. In an Address to the British Association in 1870, Maxwell made the following statement: "But a molecule, say of hydrogen, if either its mass or its time of vibration were to be altered in the least, would no longer be a molecule of hydrogen."

Our researches, however, have proved the possibility of adding new vibrations to those already existing, of changing the mean period in some cases. We may thus expect that the chemist who studies in magnetic fields the reactions taking place in luminous flames, in tubes with reduced pressure, in arc- and flame-spectra, will work as if with new molecules and atoms.

[1] Von Dechend and Hammer, "*Sitz. Ber.* Heidelberg Ak.," Abh. 21, Aug., 1910.

[2] Stark, *Physik. Zeitschr.*, **11**, 171, 1910. Stark, "Atom Dynamik," **2**.

BIBLIOGRAPHY

List of Papers by the Author concerning Radiation in Magnetic Fields.

1896.

1. Over den invloed eener magnetisatie op den aard van het door een stof uitgezonden licht I, II. (On the Influence of Magnetism on the Nature of the Light Emitted by a Substance). *Versl.* K Ak. van Wet. (Afd. Nat.), 181, Oct. 31 ; 242, Nov. 28 ; *Phil. Mag.*, 226, March, 1897 ; *Astrophys. Journ.*, **5**, No. 5, May, 1897.
2. Über einen Einfluss der Magnetsirung auf die Natur des von einer Substanz emittirten Lichtes. *Verh.* der Physik. Ges. zu Berlin, 15 Jahrg., No. 7, 128, Dec. 18.

1897.

3. L'Optique et la Théorie des Ions. *Revue Générale des Sciences pures et appliquées*, April 15.
4. Over doubletten en tripletten in het spectrum teweeggebracht door uitwendige magnetische krachten, I, II, III. (Doublets and Triplets in the Spectrum produced by External Magnetic Forces.) *Versl.* K. Ak. van Wet. (Afd. Nat.), 13, May 29 ; 99, June 26 ; 260, Oct. 30 : *Phil. Mag.*, 55, July ; 255, Sept. ; *Compt. rend.*, 1444, June 21.
5. Metingen over stralingsverschijnselen in het magnetisch veld. (Measurements concerning Radiation-Phenomena in the Magnetic Field.) *Versl.* K. Ak. van Wet. (Afd. Nat.), 408, Dec. 24 ; *Phil. Mag.*, 197, Feb., 1898.

1898.

6. On an Assymmetry in the Change of the Spectrum Lines of Iron, Radiating in a Magnetic Field. *Versl.* K. Ak. van Wet. (Afd. Nat.), 122, June 25 ; *Proc.* Roy. Acad. Amsterdam, 98, June 25 ; *Astrophys. Journ.*, **9**, Jan., 1899.

1899.

7. Waarnemingen over eene asymmetrische verandering van ijzerlijnen bij straling in een magnetisch veld. (Some Observations concerning an Asymmetrical Change of the Spectrum Lines of Iron, Radiating in a Magnetic Field.) *Versl.* K. Ak. van Wet. (Afd. Nat.), 328, Dec. 30 ; *Proc.* Roy. Acad Amsterdam, 298, Dec. 30.
8. Straling in een magnetisch veld. *Verh.* Nederl. Genees-en Natuurk. Congres, 1899, Haarlem.

1900.

9. Weiteres zur unsymmetrischen Änderung der Spectrallinien in einem Magnetfelde. *Arch. Néerl.* (2), **5**, 237.

1901.

10. Eenige waarnemingen over het oplossend vermogen van een echelonspectroscoop. (Some Observations on the Resolving Power of the Michelson Echelon Spectroscope.) *Versl.* K. Ak. van Wet. (Afd. Nat.), 298, Nov. 30; *Proc.* Roy. Acad. Amsterdam, 247, Nov. 30; *Arch. Néerl.*, Ser. II, **6**, 319, 1901.

1902.

11. Observations on the Magnetic Rotation of the Plane of Polarisation in the Interior of an Absorption Band. *Versl.* K. Ak. van Wet. (Afd. Nat.), **11**, 6, May 31; *Proc.* Roy. Acad. Amsterdam, **5**, 41; *Arch. Néerl*, Ser. II, **7**, 465; *Astrophys Journ.*, **16**, 106.
12. Osservazioni sulla rotazione magnetica del piano di polarizzazione nell' interno di una riga di assorbimento. *Rend. Linc.*, Ser. V, **11**, 1 Sem., 470, May 31.

1903.

13. Strahlung des Lichtes im magnetischen Felde. Les Prix Nobel en 1902 (May 2, 1903); *Naturw. Rundschau*, **20**, 337, 1905.
14. (With J. Geest). On the Double Refraction in a Magnetic Field near the Components of a Quadruplet. *Proc.* Roy. Acad. Amsterdam, **6**, 19, May 30; *Versl.* K. Ak. van Wet. (Afd. Nat.), **12**, 23, May 30.

1904.

15. (With J. Geest). Double Refraction near the Components of Absorption Lines Magnetically Split into Several Components. *Proc.* Roy. Acad. Amsterdam, **7**, 435; *Versl.* K. Ak. van Wet., **13**, 516, Dec. 24.

1906.

16. Recent Progress in Magneto-optics. Royal Institution of Great Britain, March 30; *Nature* **75**, 138, 160; *Le Radium*, Feb., 1907; *Naturw. Rundschau*, 22 Jahrg., No. 31, Aug. 1, 1907.
17. Magnetische splitsing der spectraallijnen en veldsterkte I, II. (Magnetic Resolution of Spectrum Lines and Magnetic Force, I, II.) *Versl.* K. Ak. van Wet. (Afd. Nat.), **14**, 838, April 27, 1906; **16**, 354, Nov. 30, 1907; *Proc.* Roy. Acad. Amsterdam, **8**, 814, April 27, 1906; **10**, 351, Nov. 30, 1907.

1907.

18. The Intensities of the Components of Spectrum Lines Divided by Magnetism. *Proc.* Roy. Acad. Amsterdam, **10**, 289, Oct. 26; *Versl.* K. Ak. van Wet. (Afd. Nat.), **16**, 286; *Le Radium*, **5**, No. 2, 49, Feb., 1908.
19. Waarneming van de magnetische splitsing der spectraallijnen met de methode van Fabry en Perot. (Observations of the Magnetic Resolution of Spectrum Lines by Means of the Method of Fabry and Perot.) *Versl.* K. Ak. van Wet. (Afd. Nat.), 486, Dec. 28; *Proc.* Roy. Acad. Amsterdam, 440, Dec. 28; *Phys. Z.S.*, 9 Jahrg., No. 7, 209.

1908.

20. Nieuwe waarnemingen over asymmetrisch door magnetisme gesplitste tripletten. (New Observations concerning Asymmetrical Triplets.) *Versl.* K. Ak. van Wet., 610, Feb. 29; *Proc.* Roy. Acad. Amsterdam, 566, Feb. 29; *Phys. Z.S.*, 9 Jahrg., No. 10, 340.
21. Verandering van golflengte van de middelste lijn van tripletten in een magnetisch veld. I, II. (Change of Wave-length of the Middle Line of Triplets, I, II.) *Versl.* K. Ak. van Wet. (Afd. Nat.), 618, Feb. 29, 855, April 24; *Proc.* Roy. Acad. Amsterdam, 574, Feb. 29, 862, April 24; *Phys. Z.S.*, 10 Jahrg., No. 7, 217; *Arch. Néerl.*, Sér. II, **14**, 267.
22. Solar Magnetic Fields and Spectrum Analysis. *Nature*, 369, Aug. 20.
23. Spektralanalytische Untersuchung der magnetischen, Felder auf der Sonne. *Verh. d. D. physik. Ges.*, 10 Jahrg., Nos. 18–20, Sept. 23; *Phys. Z.S.*, 9 Jahrg., No. 23, 834.

24. De wet van verplaatsing van de middelste lijn van een triplet. (The Law of Shift of the Central Component of a Triplet in a Magnetic Field.) *Versl.* K. Ak. van Wet. (Afd. Nat.), 541, Dec. 24; *Proc.* Roy. Acad. Amsterdam, 473, Dec. 24; *Phys. Z.S.*, 10 Jahrg., No. 7, 220; *Arch. Néerl.*, Ser. II, **14**, 273.

25. Recherches sur la décomposition magnétique des raies spectrales. *Arch. Néerl.*, Sér. II, **13**, 260.

1909.

26. Änderung der Wellenlänge an der Mittellinie von Triplets. *Phys. Z.S.*, 10 Jahrg., No. 7, 217, Feb. 22, 1909; *Arch. Néerl.*, Ser. II, **14**, 267.

27. De graad van volledigheid der circulaire polarisatie van magnetisch gesplitste lijnen. (The Degree of Completeness of the Circular Polarisation of Magnetically Divided Lines.) *Versl.* K. Ak. van Wet. (Afd. Nat.), 317, Oct. 30; *Proc.* Roy. Acad. Amsterdam, 345, Oct. 30; *Phys. Z.S.*, 11 Jahrg., SS. 1-7. *Arch. Néerl.*, Ser. II, **15**, 178.

1910.

28. (With B. Winawer). De magnetische splitsing van absorptie lijnen in verband met het spectrum der zonnevlekken. (The Magnetic Separation of Absorption Lines in Connexion with Sunspot Spectra), I, II, III. *Versl.* K. Ak. van Wet. (Afd. Nat.), 621, Jan. 29, 889, April 29, 233, June 25; *Proc.* Royal Acad. Amsterdam, 584, Jan. 29, 35, May 28, 162, June 25; *Phys. Z.S.*, 11 Jahrg., 553, 657, 880. *Astrophys. Journ.*, **32**, 329; *Arch. Néerl.*, Ser II, **15**, 453.

1911.

29. Beschouwingen over lichtstraling onder den gelijktijdigen invloed van electrische en magnetische krachten en eenige naar aanleiding daarvan genomen proeven. I. (Considerations Concerning Light Radiation under the Simultaneous Influence of Electric and Magnetic Forces and some Experiments thereby suggested. I.) *Versl.* K. Ak. 957, Jan. 28; *Proc.* Roy. Acad. Amsterdam, 2, May 27; *Phys. Zeitschr.*, 13 Jahrg., 177, 1912.

30. Le cas général de la décomposition magnétique des raies spectrales et son application en astrophysique. *Journ. de Physique,* June; *Physik Zeitschr.*, 13 Jahrg., 86, 1912.

31. (With C. M. Hoogenboom). Electrische dubbele breking in nevels. (Electric Double Refraction in some Artificial Clouds and Vapours.) I, II, III. *Versl.* K. Ak., 570, Nov. 25, 1911; 921, Jan. 27; 188, June 29, 1912; *Proc.* Roy. Acad. Amsterdam, 558, 786, 178; *Physik. Zeitschr.*, 13 Jahrg., 913, 1912.

32. Over het isoleerend vermogen van vloeibare lucht voor hooge spanning en over het electro-optisch Kerr effect in vloeibare lucht. (Note on the Insulating Power of Liquid Air for High Potentials and on the Kerr Electro-optic Effect of Liquid Air.) *Versl.* K. Ak., Dec. 30, 1911; *Proc.* Roy. Acad. Amsterdam, 650; *Physik. Zeitschr.*, 13 Jahrg., 529, 1912.

1912.

33. Over de polarisatie van het licht door de spleet van een spektroskoop en daardoor teweeggebrachte fouten. (On the Polarisation Impressed upon Light by Traversing the Slit of a Spectroscope and some Errors resulting therefrom.) *Versl.* K. Ak., Oct. 26; *Proc.* Roy. Acad. Amsterdam, 599.

1913.

34. (With R. W. Wood.) Eene methode ter waarneming van fijne absorptielijnen van metaaldampen voor onderzoekingen in sterke magnetische velden. (A Method for obtaining Narrow Absorption Lines of Metallic Vapours for Investigations in Strong Magnetic Fields). *Versl.* K. Ak., Jan. 25; *Proc.* Roy. Acad. Amsterdam.

35. De roode lithiumlijn. (The Red Lithium Line.) *Ibid.*

Bibliography of the Magnetic Resolution of Spectrum Lines.[1]

1896.

ZEEMAN. Over den invloed eener magnetisatie op den aard van het door een stof uitgezonden licht, I, II. *Versl.* K. Ak. van Wet. (Afd. Nat.), 3, Oct. 31; 242, Nov. 28. (On the Influence of Magnetism on the Nature of the Light Emitted by a Substance.) *Phil. Mag.*, 226, March, 1897; *Astrophys. Journ.*, 5, No. 5, May, 1897.

ZEEMAN. Über einen Einfluss der Magnetisirung auf die Natur des von einer Substanz emittirten Lichtes. *Verh.* der Physikal. Ges. zu Berlin, 15 Jahrg., No. 7, 128, Dec. 18.

1897.

LODGE. The Influence of a Magnetic Field on Radiation Frequency. *Proc.* Roy. Soc., **60**, Feb. 11.

LARMOR. The Influence of a Magnetic Field on Radiation Frequency. *Proc.* Roy. Soc., **60**, Feb. 11.

LODGE. The Latest Discovery in Physics. *Electrician*, 568, Feb. 26.

LODGE. A Few Notes on Zeeman's Discovery. *Electrician*, March 12.

EGOROFF AND GEORGIEWSKY. Sur la polarisation partielle des radiations émises par quelques sources lumineuses, sous l'influence du champ magnétique. *Compt. rend.*, 748, April 5; 949, May 3.

ZEEMAN. L'Optique et la Théorie des Ions. *Revue Générale des Sciences pures et appliquées*, April 15.

GARBASSO. Sul modo di interpretare certe esperienze del Sig. P. Zeeman. *Nuovo Cimento*, Series IV, **6**, April.

RICHARZ. Über die elektrischen und magnetischen Kräfte der Atome. *Elecktrot. Z.S.* May 13, Heft 19, S. 285.

E. B. F. Astrophysical Notes. *Science*, **5**, No. 124, 764, May 14.

LODGE AND DAVIES. Further Note on the Influence of a Magnetic Field on Radiation Frequency. *Proc.* Roy. Soc., **61**, 413, May 19.

ZEEMAN. Over doubletten en tripletten in het Spectrum teweeggebracht door iutwendige magnetische krachten, I, II, III. (Doublets and Triplets in the Spectrum produced by External Magnetic Forces.) *Versl.* K. Ak. van Wet. (Afd. Nat.), 13, May 29; 99, June 26; 260, Oct. 30; *Phil. Mag.*, 55, July; 255, Sept.; *Compt. rend.*, 1444, June 21.

MICHELSON. Radiation in a Magnetic Field. *Phil. Mag.*, 109, July.

POINCARÉ. La théorie de Lorentz et les expériences de Zeeman. *Eclair. electr.*, **11**, 481, June 5.

WIND. Over de dispersie der magnetische draaiing van het polarisatievlak. *Versl.* K. Ak. van Wet. (Afd. Nat.), June 26.

LORENTZ. Opmerkingen naar aanleiding van bovenstaande mededeeling (nl. C. H. WIND. Over de dispersie der magnetische draaiing van het polarisatievlak). *Versl.* K. Ak. van Wet. (Afd. Nat.), June 26.

EGOROFF AND GEORGIEWSKY. Sur la polarisation partielle des radiations lumineuses sous l'influence du champ magnétique. *Compt. rend.*, 16, July 5.

KÖNIG. Beobachtung des Zeeman'schen Phanomens. *Ann. d. Phys.*, **62**, 240, July.

KONIG. Einfache Demonstration des Zeeman'schen Phänomens. *Ann. d. Phys.*, **63**, 268, Sept.; *Eclair. électr.*, March 5, 1898.

LORENTZ. Über den Einfluss magnetischer Kräfte auf die Emission des Lichtes. *Ann. d. Phys.*, **63**, 278, Sept.; *Eclair. électr.*, March 5, 1898.

FITZGERALD. Zeeman's Phenomenon. *Nature*, 468, Sept. 16.

LORENTZ. Sur la polarisation partielle de la lumière émise par une flamme placée dans un champ magnétique. *Versl.* K. Ak. van Wet. (Afd. Nat.), 193, Sept. 25; *Arch. Néerl.*, Series II, **2**.

[1] This list is arranged chronologically. It contains so far as possible the complete literature of magnetic resolution of spectrum lines.

CORNU. Sur l'observation et l'interprétation cinématique des phénomènes découverts par M. Zeeman. *Compt. rend.*, Oct. 18 ; *J. de Phys.*, Dec. ; *Eclair. électr.*, Nov. 6 ; *Séances* Soc. Franç. de Phys., 138.

BROCA. Sur le mécanisme de la polarisation rotatoire magnétique. *Compt. rend.*, Nov. 8 ; *J. de phys.*, Dec.

H. BECQUEREL. Sur une interprétation applicable au phénomène de Faraday et au phénomène de Zeeman. *Compt. rend.*, Nov. 8 ; *J. de Phys.*, Dec.

COTTON. Procédé simple pour constater le changement de période de la lumière du sodium dans un champ magnétique. *Compt. rend.*, 865, Nov. 29.

BROCA. Les Variations de période des Raies spectrales. *Revue générale des Sciences pures et appliquées*, 935, Dec. 15.

HAMY. Sur un appareil permettant de séparer des radiations simples très voisines. *Compt. rend.*, 1092, Dec. 20.

PRESTON. Radiation Phenomena in a Strong Magnetic Field. *Trans.* Dubl. Soc. **6**, Series II, 385 (Read Dec. 22, 1897).

PRESTON. The Zeeman Effect Photographed. *Nature*, 173, Dec. 23.

ZEEMAN. Metingen over stralingsverschijnselen in het magnetisch veld. Measurements concerning Radiation-Phenomena in the Magnetic Field.) *Versl.* K. Ak. van Wet. (Afd. Nat.), 408, Dec. 24; *Phil. Mag.*, 197, Feb., 1898.

COTTON. Sur la polarisation de la lumière émise par une flamme au sodium plaçée dans un champ magnétique. *Compt. rend.*, 1169, Dec. 27.

LARMOR. On the Theory of the Magnetic Influence on Spectra; and on the Radiation from Moving Ions. *Phil. Mag.*, 503, Dec.; *Eclair. électr.*, April 9, 1898.

1898.

CORNU. Sur quelques résultats nouveaux relatifs au phénomène découvert par M. Zeeman. *Compt. rend.*, 181, Jan. 17 ; *Eclair. électr.*, Jan. 29.

H. BECQUEREL. Observations relatives à la communication de M. Cornu. *Compt. rend.*, 187, Jan. 17.

PRESTON. On the Modification of the Spectra of Iron and other Substances Radiating in a Strong Magnetic Field. *Proc.* Roy. Soc., Jan. 20 ; *Nature*, **57**, 264.

CORNU. Addition à la Note précédente sur le phénomène de Zeeman. *Compt. rend.*, 300, Jan. 24.

PRESTON On the Modification of the Spectra of Iron and other Substances Radiating in a Strong Magnetic Field. *Proc.* Roy. Soc., **63**, Jan.

MICHELSON. Radiation in a Magnetic Field. *Astrophys. Journ.*, **7**, 131, Feb. ; *Phil. Mag.*, No. 275 ; *Eclair électr.*, April 9.

HOPPE. Die Anderung der Lichtschwingungen im magnetischen Felde. *Mitth. Math. Ges.*, Hamburg, **3**, No. 8, S. 319, Feb.

COTTON. Polariseurs Circulaires. Détermination du sens d'une vibration circulaire. *J. de Phys.*, 81, Feb.

COTTON. Radiations dans un champ magnétique. *Eclair. électr.*, March 5 and 26.

LIÉNARD. La théorie de Lorentz. *Eclair. électr.*, 417, March 5 ; 456, March 12.

BROCA. Quelques propriétés des décharges électriques produites dans un champ magnétique. Assimilation au phénomène de Zeeman. *Compt. rend.*, **126**, 823, March 14 ; *Eclair. électr.*, **15**, April 2.

WIECHERT. Hypothesen für eine Theorie der elektrischen und magnetischen Erscheinungen. *Gott. Nachr.*, Heft 1, S. 87, March 19.

LORENTZ. Optische verschijnselen, die met de lading en de massa der ionen in verband staan, I, II. *Versl.* K. Ak. van Wet. (Afd. Nat.), 506, March 26 ; 555, April.

MICHELSON. On a New Spectroscope without Gratings or Prisms. *Amer. J. of Science*, March.

FITZGERALD. Note on the Connection between the Faraday Rotation of Plane of Polarisation and the Zeeman Change of Frequency of Light Vibrations in a Magnetic Field. *Proc.* Roy. Soc., **63**, 32, March.

BECQUEREL AND DESLANDRES. Contribution à l'étude du phénomène de Zeeman. *Compt. rend.*, April 4 ; *Eclair. électr.*, April 23.

CORBINO. A proposito della interpretazione del fenomeno di Zeeman data dal Sig. Cornu. *Rend. Linc.*, **7**, 1 Sem., Series V, fasc. 8, 241, April 17; *Nuovo Cimento*, Series IV, **7**, 272; *Eclair. électr.*, **15**, 548.

PRESTON. Radiation Phenomena in the Magnetic Field. *Phil. Mag.*, No. 275, April.

RIGHI. Sull' interpretazione cinematica del fenomeno di Zeeman. *Rend. Linc.*, Series V, **7**, 295, June 11; *Nuovo Cimento*, Series IV, **8**, 50.

WEISS. Un nouvel électro aimant de laboratoire donnant un champ de 30,000 unités. *Eclair. électr.*, **15**, 481, June 18.

LORENTZ. Beschouwingen over den invloed van een magnetisch veld op de uitstraling van licht. (Considerations concerning the Influence of a Magnetic Field on the Radiation of Light. *Versl.* K. Ak. van Wet (Afd. Nat.), 113, June 25; *Proc.* Roy. Acad. Amsterdam, 90, June 25.

ZEEMAN. On an Assymmetry in the Change of the Spectrum Lines of Iron, Radiating in a Magnetic Field. *Versl.* K. Ak. van Wet (Afd. Nat.), 122, June 25; *Proc.* Roy. Acad. Amsterdam, 98, June 25; *Astrophys. Journ.*, **9**, Jan., 1899.

MICHELSON. The Echelon Spectroscope. *Astrophys. Journ.*, **8**, 37, June.

AMES, EARHART, REESE. Notes on the Zeeman Effect. *Astrophys. Journ.*, **8**, 48, June.

AMES, EARHART, REESE. Some Notes on the Zeeman Effect. Johns Hopkins University *Circulars*, **17**, No. 135, June.

BECQUEREL AND DESLANDRES. Nouvelles Observations sur le phénomène de Zeeman. *Compt. rend.*, 18, July 4.

RIGHI. Di un nuovo metodo sperimentale per lo studio dell' assorbimento della luce nel campo magnetico. *Rend. Linc.*, **7**, 2 Sem., Series V, July 17; *Compt. rend.*, 216, July 25; *Berl. Sitzber.*, July 28.

v. AUBEL. Action du magnétisme sur les spectres des gaz. *J. de Phys.*, July.

PRESTON. On Radiation from a Source of Light in a Magnetic Field. *Rep.* Brit. Assoc., 789, Sept. 9.

SILVANUS THOMPSON. On the Discovery by Righi of the Absorption of Light in a Magnetic Field. *Rep.* Brit. Assoc., Sept. 9, 789.

MACALUSO AND CORBINO. Sur une nouvelle action subie par la lumière traversant certaines vapeurs métalliques dans un champ magnétique. *Compt. rend.*, 548, Oct. 17.

BECQUEREL. Remarques sur la polarisation rotatoire magnétique et la dispersion anomale, à l'occasion d'une expérience nouvelle de MM. Macaluso et Corbino. *Compt. rend.*, 647, Oct. 31.

VOIGT. Über den Zusammenhang zwischen dem Zeeman'schen und dem Faraday 'schen Phänomen. *Gött. Nachr.*, Heft 4, 1, Oct.; *Verhandl. d. Gesellsch. deutscher Naturf.* Düsseldorf, 43.

FITZGERALD. The Zeeman Effect and Dispersion. *Science Progress*, Oct.

VOIGT. Zur Theorie der magneto-optischen Erscheinungen. *Ann. d. Phys.* **67**, S. 345, Nov.

VOIGT. Zur Theorie der von den Herren Macaluso und Corbino entdeckten Erscheinungen. Doppelbrechung von im Magnetfelde befindlichem Natrium dampf in der Richtung normal zu den Kraftlinien. *Gött. Nachr.*, Heft 4, Nov.

MACALUSO AND CORBINO. Sopra una nuova azione che la luce subisce attraversando alcuni vapori metallici in un campo magnetico. *Rend. Linc.*, **7**, 293, Nov. 20; *Compt. rend.*, 951, Dec. 5.

BECQUEREL. Sur la dispersion anomale et le pouvoir rotatoire magnétique de certaines vapeurs incandescentes. *Compt. rend.*, 899, Dec. 5.

COTTON. Absorption dans un champ magnétique. *Compt. rend.*, 953, Dec. 5.

BECQUEREL. Observations au sujet de la Communication précédente. *Compt. rend.*, 953, Dec. 5.

VOIGT. Über die Proportionalität von Emissions und Absorptions-vermögen. *Ann. d. Phys*, **67**, S. 366, Dec.

1899.

RIGHI. Sur l'absorption de la lumière par un corps placé dans un champ magnétique. *Compt. rend.*, 45, Jan. 2.

FITZGERALD. Converse of the Zeeman Effect. *Nature*, 222, Jan. 5.
PRESTON. Radiation Phenomena in the Magnetic Field. *Nature*, 224, Jan. 5.
MACALUSO AND CORBINO. Sulle modificazione che la luce subisce attraversando alcuni vapori metallici in un campo magnetico. *Rend. Linc.*, 8, 1 Sem., Series V^a^, fasc. 1, Jan. 8.
PRESTON. General Law of Magnetic Perturbations of Spectrum Lines. *Nature*, 248, Jan. 12.
BECQUEREL. Sur la dispersion anomale de la vapeur de sodium incandescente, et sur quelques conséquences de ce phénomène. *Compt. rend.*, 146, Jan. 16.
PRESTON. Radiating Phenomena in a Strong Magnetic Field, Part II, Magnetic Perturbations of the Spectrum Lines. *Trans.* Dubl. Soc., **7**, Series II, 7 (Read Jan. 18, 1899).
STONEY, PRESTON. Illusory Resolutions of the Lines of a Spectrum. *Nature*, 294, Jan. 26.
LORENTZ. Trillingen van electrisch geladen stelsels in een magnetisch veld. (Sur les vibrations de systèmes portant des charges électriques et placés dans un champ magnétique.) *Versl.* K. Ak. van Wet. (Afd. Nat.), 320, Jan. 28; *Arch. Néerl.*, Series II, **2**; *Proc.* Roy. Acad. Amsterdam, 340, Jan. 28.
COTTON. Biréfringence produite par le champ magnétique, liée au phénomène de Zeeman. *Comp. rend.*, 294, Jan. 30; *Eclair. électr.*, **18**, 520.
GOLDHAMMER. Das Zeeman'sche Phänomen, die magnetische Circularpolarisation und die magnetische Doppelbrechung. *Ann. d. Phys.*, **67**, S. 697, Jan.
MACALUSO AND CORBINO. Sulla relazione tra il fenomeno di Zeeman e la rotazione magnetica anomala del piano di polarizzazione della luce. *Rend. Linc.*, **8**, 1 Sem., Series V, fasc. 3, Feb. 5.
COTTON. L'aspect actuel de la loi de Kirchhoff. *Revue Générale des Sciences pures et appliquées*, Feb. 15.
GRAY. Production of Magnetisation by Circularly Polarised Light. *Nature*, 367, Feb. 16.
PRESTON. Magnetic Perturbations of the Spectrum Lines. Further Resolution of the Quartet. *Nature*, 367, Feb. 16.
CORBINO. Sui battimenti luminosi e sull' impossibilità di produrli ricorrendo al fenomeno di Zeeman. *Rend Linc.*, **8**, 1 Sem., Series V, fasc. 4, Feb. 19.
PRESTON. Radiation Phenomena in the Magnetic Field. Magnetic Perturbations of the Spectrum Lines. *Phil. Mag.*, Feb., No. 285.
CORBINO. Sulla dipendenza tra il fenomeno di Zeeman e le altre modificazioni che la luce subisce dai vapori metallici in un campo magnetico. *Rend. Linc.*, **8**, 1 Sem., Series V, fasc. 5, March 5.
LARMOR. On the Origin of Magnetic-optic Rotation. *Cambridge Phil Soc.*, March 6; *Nature*, 527, March 30.
LARMOR. Remarks on the Phenomenon of Zeeman and its Bearing on the Problem of the Origin of Spectra. *Math. Soc.*, March 9.
MICHELSON. Radiation in a Magnetic Field. *Nature*, 440, March 9.
PRESTON. Radiation in a Magnetic Field. *Nature*, 485, March 23.
RIGHI. Sul potere rotatorio magnetico del cloro. *Rend. di Bol.*, March 26.
FITZGERALD. Experiment to Illustrate the Zeeman Effect. *Nature*, 509, March 30.
RAYLEIGH. The Interferometer. *Nature*, 533, April 6.
POINCARÉ. La Théorie de Lorentz et le phénomène de Zeeman. *Eclair. électr.*, **19**, 5, April 8.
RIGHI. Intorno alla questione della produzione di un campo magnetico, per opera di un raggio luminoso polarizzato circolarmente. *Rend. Linc.*, **8**, 1 Sem., Series V, fasc. 7, April 9.
PRESTON. The Interferometer. *Nature*, 605, April 27.
BUTLER. The Michelson Echelon Spectroscope. *Nature*, 606, April 27.
DRUDE. Zur Theorie der magneto-optischen Erscheinungen. *Verh d. D. physik. Ges.*, 1 Jahrg., No. 6, April.
PRESTON. Magnetic Perturbations of the Spectrum Lines. Royal Institution of Great Britain, May 12.
VOIGT. Weiteres zur Theorie des Zeeman Effektes. *Ann. d. Phys.*, **68**, S. 352, May.

DE HEMPTINNE. Influence du magnétisme sur la luminosité des gaz. *Bull.* de Belg., Part 6, 447, June 3.

PRESTON. Magnetic Perturbations of the Spectrum Lines. *Nature*, **60**, 175, June 22.

LORENTZ. La théorie élémentaire du phénomène de Zeeman. Réponse à une objection de M. Poincaré. *Versl.* K. Ak. van Wet. (Afd. Nat.), **8**, 69, June 24; *Proc.* Roy. Acad. Amsterdam, **2**, 52; *Arch. Néerl.*, Series II, **7**, 299, 1902.

REESE. Notes on the Zeeman Effect. Johns Hopkins University *Circulars*, No. 140, June.

MICHELSON. Sur le spectroscope à échelons. *J. de phys.*, 305, June.

SHEDD. An Interferometer Study of Radiation in a Magnetic Field. *The Phys. Rev.*, **9**, No. 1, July; No. 2, Aug.

RIGHI. Sull' assorbimento della luce per parte di un gas posto nel campo magnetico. *Nuovo Cimento.*, Series IV, **10**, 20, July.

VOIGT. Zur Theorie der Einwirkung eines elektrostatischen Feldes auf die optischen Eigenschaften der Korper. *Ann. d. Phys.*, **69**, S. 297, Aug. 15.

LORD KELVIN. Magnetism and Molecular Rotation. *Phil. Mag.* Aug., No. 291.

VOIGT. Bemerkung uber die bei dem Zeeman'schen Phänomen stattfindenden Intensitätsverhältnisse. *Ann. d. Phys.*, **69**, Aug., S. 290.

LORENTZ. Zur Theorie des Zeemaneffektes. *Phys. Z.S.*, Sept. 4, 1 Jahrg., No. 3.

MICHELSON. The Echelon Spectroscope. *Trans.* Cambr. Soc., 1900, 316 (Received Oct. 19, 1899).

VOIGT. Neuere Untersuchungen über die optischen Wirkungen eines Magnetfeldes. *Phys. Z.S.*, 1 Jahrg., Nos. 10, 11, 12, Oct. 21.

PRESTON. Some Remarks on Radiation Phenomena in a Magnetic Field. *Nature* 11, Nov. 2.

CORBINO. Reciprocita nei fenomeni magneto-ottici. *Nuovo Cimento*, Series IV **10**, Nov.

RIGHI. Sul fenomeno di Zeeman nel caso generale d'un raggio luminoso comunque inclinato sulla direzione della forza magnetica. *Mem. di Bol.*, Dec. 17.

VOIGT. Über eine Dissymmetrie der Zeeman'schen normalen Triplets. *Ann. d. Phys.*, **1**, S. 376 (Eingegangen Dec. 23, 1899); *Eclair. électr.*, May 26, 1900.

VOIGT. Weiteres zur Theorie der magneto-optischen Wirkungen. *Ann. d. Phys.*, **1**, S. 389 (Eingegangen Dec. 23, 1899).

ZEEMAN. Waarnemingen over eene asymmetrische verandering van ijzerlijnen bij straling in een magnetisch veld. (Some Observations concerning an Asymmetrical Change of the Spectrum Lines of Iron, Radiating in a Magnetic Field) *Versl.* K. Ak. van Wet. (Afd. Nat.), 328, Dec. 30; *Proc.* Roy. Acad. Amsterdam, 298, Dec. 30.

ZEEMAN. Straling in een magnetisch veld. *Verh.* Nederl. Genees-en Natuurk. Congres, 1899, Haarlem.

COTTON. Le phénomène de Zeeman. Scientia Series, No. 5, Gauthier-Villars, Paris, 1899.

1900.

LARMOR. On the Dynamics of a System of Electrons or Ions; and on the Influence of a Magnetic Field on Optical Phenomena. *Trans.* Cambr. Soc., **18**, 380, Jan. 24.

DONGIER. Lumière polarisée émise par un tube de Geissler soumis à l'action d'un champ magnétique. *Compt. rend.*, 244, Jan. 29; *Eclair. électr.*, **22**, 238

SHEDD. Untersuchung der Strahlung in einem magnetischen Felde mittelst des Interferometers. *Phys Z.S.*, S. 270, 1 Jahrg., No. 24, March 4.

DONGIER. Dissymétrie dans l'émission polarisée d'un tube de Geissler soumis à l'action d'un champ magnétique. *Compt. rend.*, 650, March 5.

EGOROFF AND GEORGIEWSKY. Sur les particularités optiques des tubes de Geissler sous l'influence d'un champ magnétique. *Compt. rend.*, 900, April 2.

RIGHI. Über das Zeemansche Phanomen in dem allgemeinen Falle eines beliebig gegen die Richtung der magnetischen Kraft geneigten Lichtstrahles. *Phys. Z.S.*, April 28, 1 Jahrg., No. 30, S. 329.

LORD BLYTHSWOOD AND MARCHANT. The Echelon Spectroscope, with Application to Investigate the Behaviour of the Chief Lines in the Mercury Spectrum under the Influence of a Magnetic Field. *Phil. Mag.*, (5) **49**, 384, April.

LARMOR. Magnetic Influences on Radiation as a Clue to Molecular Constitutions. "Aether and Matter," Appendix F, 341, May, 1900.

REESE. Notes on the Zeeman Effect. Johns Hopkins University *Circulars*, No. 146, June; *Electrical World*, 248, Aug. 11.

RUNGE AND PASCHEN. Studium des Zeeman-effektes im Quecksilberspektrum. *Phys. Z.S.*, No. 44, 480, 1 Jahrg., July 24.

PASCHEN. Eine Geissler'sche Rohre mit Quecksilberelektroden zum Studium des Zeemaneffektes. *Phys. Z.S.*, No. 44, 478, 1 Jahrg., July 24.

LORENTZ. Théorie des Phénomènes Magnéto-optiques. Rapports présentés au Congrès International de Physique, réuni à Paris en 1900, Aug. 6–12.

REESE. An Investigation of the Zeeman Effect. *Astrophys. Journ.*, 120, Sept.

VOIGT. Über das elektrische Analogon des Zeeman-effektes. *Ann. d. Phys.*, **4**, S. 197, Nov. 13, 1900.

SHEDD. Die Anwendung des Interferometers beim Studium des Zeeman effektes. *Phys. Z.S.*, **2**, S. 278, No. 19, Dec. 16, 1900; Colorado Springs, Colo. Coll. Stud., *Sci. Ser.*, No. 38, 106, 1904.

LEWIS. Effects of a Magnetic Field on Radiation. Memoirs by Faraday, Kerr, Zeeman.

ZEEMAN. Weiteres zur unsymmetrischen Änderung der Spectrallinien in einem Magnetfelde. *Arch. Néerl.*, (2), **5**, 237.

BLUMENTHAL. Die Bewegung der Ionen beim Zeeman'schen Phänomen. *Z.S. f. Math. u. Phys.*, **45**, Nos. 2–3.

1901.

KAUFMANN. Über die Schwingungsamplitude der Elektronen. *Phys. Z.S.*, **2**, No. 19, S. 283, Jan. 20.

REESE. Der Zeeman-Effekt. *Phys. Z.S.*, **2**, No. 25, S. 369, Feb. 9.

ZEHNDER. Über Gitterbeobachtungen. *Ann. d. Phys.*, **5**, S. 685, April.

KENT. Notes on the Zeeman Effect. Johns Hopkins University *Circulars*, May–June; *Phil. Mag.*, Sept.; *Astrophys. Journ.*, **13**, No. 5.

VOIGT. Beiträge zur Elektronentheorie des Lichtes. *Ann. d. Phys.*, **6**, Aug.

CORBINO. Dispersione rotatoria magnetica dei vapori di sodio nell' interno della riga di assorbimento. *Rend. Linc.*, **10**, 2 Sem., Series V, fasc. 6, Sept. 15.

KAUFMANN. Die Entwicklung des Elektronenbegriffs. *Phys. Z.S.*, No 1, S. 9–15, Sept.; *Naturw. Rundschau*, **16**, 557, 569.

LECHER. Die Hertz'schen Entdeckung elektrischer Wellen und deren weitere Ausgestaltung. *Die Umschau*, No. 142, Oct. 12.

CORBINO. Sulla doppia rifrazione circolare e la polarizzazione rotatoria. *Rend. Linc.*, **10**, 2 Sem., Series V, fasc. 8, Oct. 20.

WALKER. An Asymmetry of the Zeeman Effect. *Nature*, 668; *Phys. Soc.*, Oct. 25.

VOIGT. Magnetische Drehung der Polarisationsebene innerhalb eines Absorptionsstreifens. *Ann. d. Phys.*, **6**, S. 784, Oct.

VOIGT. Electronenhypothese und Theorie des Magnetismus. *Gött. Nachr.*, Heft 3, Nov. 8; *Ann. d. Phys.*, **9**, 115, 1902.

GRAY AND STEWART. Effect of a Magnetic Field on Spectra of Helium and Mercury. *Nature*, 54, Nov. 21.

LORD BLYTHSWOOD AND ALLEN. The Red (C) Line of Hydrogen and the Zeeman Effect. *Nature*, 79, Nov. 28.

ZEEMAN. Eenige waarnemingen over het oplossend vermogen van een echelon-spectroscoop. (Some Observations on the Resolving Power of the Michelson Echelon Spectroscope.) *Versl.* K. Ak. van Wet. (Afd. Nat.), 298, Nov. 30; *Proc.* Roy. Acad. Amsterdam, 247, Nov. 30; *Arch. Néerl.*, Ser. III, **6**, 319, 1901.

JEANS. Mechanism of Radiation. The Zeeman Phenomenon. *Phil. Mag.*, **2**, No. 11, 441, Nov.

ZILOV. Le phénomène de Zeeman. *Fiz. Obozr. Varsăva*, **2**, 284.

1902.

LUMMER AND GEHRCKE. Über den Bau der Quecksilberlinien ; ein Beitrag zur Auflösung feinster Spectrallinien. *Berl. Sitzber.*, Jan. 9.

ABRAHAM. Dynamik des Elektrons. *Gött. Nachr.*, Heft 1, Jan. 11.

DRUDE. Zur Elektronentheorie der Metalle. *Ann d. Phys.*, **7**, S. 687, Jan.

RUNGE AND PASCHEN. Über die Strahlung des Quecksilbers im magnetischen Felde. *Abh. d. Berl. Ak.* (Anhang), Feb. 6 ; *Astrophys. Journ.*, **15**, 235.

DU BOIS. Zur magnetischen Rotationspolarisation absorbirender Substanzen. *Ann. d. Phys.* (4 Folge), **7**, 944. Feb. 25.

CORBINO. Nuove ricerche sulla polarizzazione rotatoria magnetica nell' interno di una riga d'assorbimento. *Nuovo Cimento*, Series V, **3**, Feb.

WALKER. On Asymmetry of the Zeeman Effect. *Phil. Mag.*, No. 14 (6), **3**, 247, Feb.

KORN AND STOECKL. Studien zur Theorie der Lichterscheinungen I. Das Zeeman'sche Phänomen. *Ann. d. Phys.*, **8**, S. 312, March.

RUNGE AND PASCHEN. Über die Zerlegung einander entsprechender Serienlinien im magnetischen Felde. *Berl. Sitzber.*, 380, April 10 ; 720, June 26 ; *Astrophys. Journ.*, **15**, 333 ; **16**, 123.

BERNDT. Gasspectra im Magnetfelde. *Ann. d. Phys.*, (4), **8**, S. 625, April 16.

VOIGT. Dispersione rotatoria magnetica nell'interno delle righe d'assorbimento. *Rend. Linc.*, Series V, **11**, 1 Sem., 459, May 31.

ZEEMAN. Observations on the Magnetic Rotation of the Plane of Polarisation in the Interior of an Absorption Band. *Versl.* K. Ak. van Wet. (Afd. Nat.), **11**, 6, May 31 ; *Proc.* Roy. Acad. Amsterdam, **5**, 41 ; *Arch Néerl.*, Series II, **7**, 465 ; *Astrophys. Journ.*, **16**, 106.

ZEEMAN. Osservazioni sulla rotazione magnetica del piano di polarizzazione nell' interno di una riga di assorbimento. *Rend. Linc.*, Series V, **11**, 1 Sem., 470, May 31.

RIECKE. Zeeman Effekt und Elektronenladung. *Phys. Z.S.* (**3**), No. 18, S. 406, May 31.

SCHMAUSS. Magnetische Drehung der Polarisationsebene innerhalb eines Absorptionsstreifens. *Ann. d. Phys.*, **8**, 842, May.

VOIGT. Sul fenomeno, Majorana. *Rend. Linc.*, Series V, **11**, 1 Sem., 505, June 15 ; *Nuovo Cimento*, Series V, **4**, 52, 1902.

VOIGT. Über einige neuere Beobachtungen von magneto-optischen Wirkungen. *Ann. d. Phys.* (4), **8**, 872, June.

RUNGE. Über den Zeemaneffekt der Serienlinien. *Phys. Z.S.*, S. 441, June.

MAJORANA. Sur la biréfringence magnétique. *Compt. rend.* **135**, 159, July 21.

VOIGT. Neue Beobachtungen uber magneto-optische Erscheinungen in Absorptionsstreifen. *Gött. Nachr.*, Heft 5, July.

KORN AND STOECKL. Studien zur Theorie der Lichterscheinungen II. Die Drehung der Polarisationsebene im magnetischen Felde. *Ann. d. Phys.*, **9**, 1138, Aug. 17.

FÄRBER. Uber das Zeeman-Phänomen. *Ann. d. Phys.*, 4 Folge, **9**, 886, Aug. 31 ; *Inaug. Diss. Tübingen.*

CORBINO. Sull' ineguale assorbimento della vibrazioni circolari inverse per il passaggio attraverso a un vapore incandescente in un campo magnetico. *Rend. Linc.*, Oct. ; *Nuovo Cimento*, Series V, **6**, 58 ; 1903 ; *Giornale* della Società di scienze naturali ed economiche, Palermo, **25**, 1905, 9.

CORBINO. La rotazione magnetica del piano di polarizzazione nell' interno di una riga d'assorbimento. *Rend. Linc.*, Oct. ; *Nuovo Cimento*, Series V, **6**, 55, 1903 ; *Giornale* della Società di scienze naturali ed economiche, Palermo, **25**, 13, 1905.

MAJORANA. Quelques phénomènes magnéto-optiques présentés par des solutions magnétiques. *Verh. d. schweiz. Naturf. Ges.*, **85**, 47 ; *Arch. sci. phys.*, Series IV, **14**, 361, Oct.

LODGE. Electrons. *J. Inst. Electr. Engin.*, Part 159, **32**, Nov. 27.

SCHMAUSS. Magnetische Drehung der Polarisationsebene des Lichles in selektiv absorbirenden Medien. *Sitzber. d. Bayer. Akad.*, **32**, Heft 3, Nov. ; *Ann. d. Phys.*, **10**, 853, Dec. 28.

LORENTZ. Nobel Vorlesung, Dec. 11 ; Les Prix Nobel en 1902.

SCHMAUSS. Notiz zur magnetischen Doppelbrechung. *Ann. d. Phys.*, **10**, 658, Dec. 31, 1902.

HALLO. De magnetische draaiing van het polarisatievlak in de nabijheid van een absorptieband. *Diss. Amsterdam*, 1902; *Arch. Néerl.*, Serie II, **10**, 148, 1905.

FRIEDRICH. Die elektrischen Atome und die spezifische Ladung der Ionen. *Wissenschaftliche Beilage zum Jahresbericht* des Städt. Gymn. i. E. mit Realschule zu Sölingen.

J. J. THOMSON. Magneto-optics. Supplement to the *Encycl. Brit.*, **30**, 464.

1903.

DE HEEN. Oscillations de l' éther de diverses formes et le phénomène de Zeeman. *Bull.* de Belg. No. 1, 41, Jan. 10.

v. BACKLUND. Uber die magneto-optischen Erscheinungen. *Arkiv för Matematik, Astronomi och Fysik*, **1**, Jan. 14.

HALLO. The Value of Some Magneto-Optic Constants. *Versl.* K. Ak. van Wet. (Afd. Nat.), **11**, 535, Jan. 31; *Proc.* Roy. Acad. Amsterdam, **5**, 438.

RUNGE AND PRECHT. Die Stellung des Radium im periodischen System nach seinem Spektrum. *Phys. Z.S.*; S. 285, Feb.

v. D. WAALS, SR. Het Zeeman-Verschijnsel. *De Gids*, March.

HALLO. Rotation magnétique du plan de polarisation et absorption sélective. *Handl.* Ned. Nat. Geneesk. Congres, **9**, 126, April.

ZEEMAN. Strahlung des Lichtes im magnetischen Felde. Les Prix Nobel en, 1902 (May 2, 1903); *Naturw. Rundschau*, **20**, 337, 1905.

GRAY, STEWART, HOUSTOUN, MCQUISTON. On the Radiation of Helium and Mercury in a Magnetic Field. *Proc.* Roy. Soc., **72**, 16, May 14.

ZEEMAN AND GEEST. On the Double Refraction in a Magnetic Field near the Components of a Quadruplet. *Proc.* Roy. Acad. Amsterdam, **6**, 19, May 30; *Versl.* K. Ak. van Wet. (Afd. Nat), **12**, 23, May 30.

RUNGE. Über die spektroskopische Bestimmung des Atomgewichtes. *Verh. d. D. physik. Ges.*, S. 313; *Phys. Z.S.*, S. 752, Oct.

DESLANDRES. Caractères principaux des spectres de lignes et de bandes. Considérations sur les origines de ces deux spectres. *Compt. rend.*, 24, **137**, 1013, Dec. 14.

SIERTSEMA. The Calculation of e/m from the Magnetic Rotations of the Plane of Polarisation for Substances without an Absorption Band in the Visible Spectrum. *Proc* Roy. Acad. Amsterdam, **5**, 413, Dec. 27; *Versl.* K. Ak. van Wet. (Afd. Nat.), **11**, 499; Commun. Lab. of Phys. Leiden, No. 82.

J. J. THOMSON. The Magnetic Properties of Systems of Corpuscles describing Circular Orbits. *Phil. Mag.*, Dec., No. 36, 673.

JAGER. Das Zeemanphänomen. *Schriften* des Vereins zur Verbreitung Naturwissenschaftlicher Kenntnisse in Wien, **43**, 213.

1904.

HEAVISIDE. The Radiation from an Electron Describing a Circular Orbit. *Nature*, **69**, 293, Jan. 28.

HEAVISIDE. The Radiation from an Electron Moving in an Elliptic, or any other Orbit. *Nature*, **69**, 342, Feb. 11.

RUNGE AND PRECHT. Über die magnetische Zerlegung der Radiumlinien. *Berl. Sitzber.*, Feb. 18.

ROBB. Beiträge zur Theorie des Zeeman effektes. *Inaug. Diss. Göttingen*, Feb.; *Ann. d. Phys.*, **15**, 107.

J. J. THOMSON. Note on the Effect of a Magnetic Field on the Vibrations of an Atom containing Six Corpuscles placed at the Corners of a Regular Octahedron. *Cambridge Phil. Soc.*, May 17.

NAGAOKA. A Dynamical System Illustrating the Spectrum Lines. *Nature*, June 9, 124; *Phil. Mag.*, (6), **7**, 445, May.

NUTTING. Atomic Structure in the Light of Secondary Spectra. *Nature*, 342, Aug. 11.

WOOD. A Quantitative Determination of the Anomalous Dispersion of Sodium Vapour in the Visible and Ultraviolet Regions. *Phil. Mag.*, **8**, 301, Sept.

LORENTZ. Ergebnisse und Probleme der Elektronentheorie Vortrag gehalten am Dec. 20, 1904, im Elektrotechn. Verein zu Berlin. *Verlag von J. Springer*, Berlin, 1905.

ZEEMAN AND GEEST. Double Refraction near the Components of Absorption Lines Magnetically Split into Several Components *Proc.* Roy. Acad. Amsterdam, **7**, 435 ; *Versl.* K. Ak. van Wet , **13**, 516, Dec. 24.

J. J. THOMSON. On the Vibrations of Atoms containing 4, 5, 6, 7, and 8 Corpuscles, and on the Effect of a Magnetic Field on such Vibrations (Received Dec. 27.) *Proc. Camb. Phil. Soc.*, **13**, 39, 1906.

GEEST. La double réfraction magnétique de la vapeur de sodium. *Diss.* Amsterdam, 1904 ; *Arch. Néerl.*, Series II, **10**, 291, 1905.

SCHUSTER. An Introduction to the Theory of Optics, 167.

1905.

LANGEVIN. Sur la theorie du magnétisme. *Soc. Franç. de Phys.*, Feb. 3.

PURVIS. The Influence of Strong Electro-magnetic Fields on the Spark Spectra of some Metals. *Cambridge Phil. Soc.*, Feb 27 ; *Nature*, 479, March 16.

PURVIS. The Influence of Very Strong Electro-magnetic Fields on the Spark Spectra of Gold, Antimony, Bismuth, Lead and Tin. *Proc. Camb. Soc.*, **13**, 82, Feb. 27.

GEEST. Uber die Doppelbrechung von Natriumdampf im magnetischen Felde, *Phys. Z.S.*, **6**, 166, Feb.

WOOD. Anomale Dispersion der magnetischen Drehung der Polarisationsebene, *Phys. Z.S.*, **6**, 416, No. 13, April 20.

WOOD. The Magneto-optics of Sodium Vapour and the Rotatory Dispersion Formula. *Proc.* Phys. Soc., **19**, 742, June 30 ; *Phil. Mag.*, Series VI, **10**, 408, Oct.

WOOD AND SPRINGSTEEN. The Magnetic Rotation of Sodium Vapour. *The Phys. Rev.*, **21**, 41, July.

COTTON AND MOUTON. Biréfringence magnétique. Nouveaux liquides actifs, *Compt. rend.*, **141**, 349, Aug. 7.

PURVIS. The Influence of Very Strong Magnetic Fields on the Spark Spectra of Ruthenium, Rhodium and Palladium. *Nature*, Nov. 16, No. 1881 ; *Cambridge Phil. Soc.*, **13**.

LORENTZ. De absorptie en emissiebanden van gasvormige lichamen I and II (The Absorption and Emission Lines of Gaseous Bodies). *Versl.* K. Ak. van Wet. (Afd. Nat.), Nov. and Dec. ; *Proc.* Roy. Acad. Amsterdam, **8**, 591.

JANICKI. Feinere Zerlegung der Spektrallinien von Quecksilber, Kadmium, Natrium, Zink, Thallium und Wasserstoff, *Inaug Diss.*, Halle.

AUERBACH. Beziehungen des Magnetismus zum Licht. Handbuch der Physik, **5**, S. 374

1906.

PURVIS. The Influence of Very Strong Electromagnetic Fields on the Spark Spectra of (1) Vanadium and (2) Platinum and Iridium. *Trans. Camb. Phil. Soc.*, **20**, Feb. ; *Nature*, **73**, 527.

J. BECQUEREL. Sur les variations des bandes d'absorption d'un cristal dans un champ magnétique, *Compt. rend.*, March 26, April 9.

ZEEMAN. Recent Progress in Magneto-optics. Royal Institution of Great Britain, March 30 ; *Nature*, **75**, 138, 160 ; *Le Radium*, Feb., 1907 ; *Naturw. Rundschau*, 22 Jahrg., No. 31, Aug. 1, 1907.

ZEEMAN. Magnetische splitsing der spectraallijnen en veldsterkte, I en II (Magnetic Resolution of Spectrum Lines and Magnetic Force, I, II). *Versl.* K. Ak. van Wet. (Afd. Nat.), **14**, 838, April 27, 1906 ; **16**, 354, Nov. 30, 1907 ; *Proc.* Roy. Acad. Amsterdam, **8**, 814, April 27, 1906 ; **10**, 351, Nov. 30, 1907.

COTTON AND MOUTON. Aimants transparents. Propriétés optiques et magnetiques. *Bull. Soc. Franç. de Phys.*, 49, May 4.

PURVIS. The Influence of a Very Strong Magnetic Field on the Spark Spectra of Palladium, Rhodium, and Ruthenium. *Camb. Phil. Soc.*, **13**, Pt. 6, 325, May 14.

PURVIS. Experiments on the Band Spectrum of Nitrogen in a Strong Magnetic Field. *Camb. Phil. Soc.*, **13**, Pt. 6, 354, May 14.

J. BECQUEREL. Sur la corrélation entre les variations des bandes d'absorption des cristaux dans un champ magnétique et la polarisation rotatoire magnétique. *Compt. rend.*, **142**, 1144, May 21.

VOIGT. Fragen der Kristallphysik II. Über die Wirkung eines Magnetfeldes auf das optische Verhalten pleochroitischer Kristalle. *Gött. Nachr.*, 507 July 28.

LOHMANN. Beiträge zur Kenntniss des Zeeman-Phänomens. *Phys. Z.S.*, No. 22, S. 809, Oct.; *Inaug. Diss.*, Halle, 1907.

GEHRCKE AND V. BAYER. Über den Zeeman-Effekt in schwachen Magnetfeldern, *Verh. d. D. physik Ges.*, 8 Jahrg., No. 20, Sept. 18; *Phys. Z.S.*, 7 Jahrg, S. 905, No. 24, Oct. 9.

MILLER. Zeeman Effekt an Mangan und Chrom. *Phys. Z.S.*, 7 Jahrg., S. 896, No 24, Nov. 10.

J. BECQUEREL. Sur une théorie des phénomènes magnéto-optiques dans les cristaux. *Compt. rend.*, Nov. 19.

PURVIS. The Influence of a Strong Magnetic Field on the Spark Spectra of Titanium, Chromium, and Manganese. *Camb. Phil. Soc.*, Nov. 26; *Nature*, **75**, 189, Dec. 20.

DUFOUR. Sur les spectres de l'hydrogène. Action d'un champ magnétique. Phénomène de Zeeman. *Ann. chim. phys*, **9**, Nov.

J. BECQUEREL. Sur une explication théorique des phénomènes magnéto-optiques observés dans un cristal. *Compt. rend.*, Dec. 2.

J. BECQUEREL. Sur les phénomènes de polarisation rotatoire magnétique dans les cristaux. *Compt. rend.*, Dec. 10.

J. BECQUEREL. Sur les modifications dissymétriques de quelques bandes d'absorption d'un cristal sous l'action d'un champ magnétique. *Compt. rend.*, **143**, 1133, Dec. 24.

WOOD. Fluorescence and Magnetic Rotation Spectra of Sodium Vapour and their Analysis. *Proc.* Phys. Soc, **20**, 345.

FEFELOV. Les phénomènes magnéto-optiques (Russe). *Věst. opytn. fiziki*, Odessa, **35**, 103, 126, 145.

BIERNACKI. Sur le phénomène de Zeeman. *Przegl.-techn.*, **44**, 137, 157, 177, 211, 242, 266, 296, 322.

1907.

J. BECQUEREL. Sur les phenomènes magnéto-optiques dans les cristaux. *Soc. Franç. de Phys.*, No. 255, **5**, Jan. 18; *Le Radium*, Feb.

J. BECQUEREL. Sur les modifications subies dans un champ magnétique par les bandes d'absorption des cristaux de tysonite. *Compt. rend.*, **144**, 132, Jan. 21.

WEISS AND COTTON. Mesures du phenomène de Zeeman sur les raies bleues du zinc. *Compt. rend.*, Jan. 21; *Soc. Franç. de Phys.*, April 19; *J. de Phys.*, (4), **6**, 429, June.

WOOD. Ionisation and Anomalous Dispersion. *Nature*, **75**, 390, Feb. 21.

J. BECQUEREL. Influence de la température sur l'absorption dans les cristaux. Phénomènes magnéto-optiques à la température de l'air liquide. *Compt. rend.*, **144**, 420, Feb. 25.

SCHOTT. Ionisation and Anomalous Dispersion. *Nature*, **75**, 461, March 14.

J. BECQUEREL. Sur la polarisation rotatoire magnétique aux environs des bandes d'absorption. Pouvoir rotatoire magnétique des cristaux à la température de l'air liquide. *Compt. rend.*, **144**, 682, March 25.

PAGE. The Variation of the Absorption Bands of a Crystal in a Magnetic Field. *Trans. Cambr. Soc.*, **20**, 291, March 30.

RUNGE. Über die Zerlegung von Spektrallinien im magnetischen Felde. *Phys. Z.S.*, **8**, S. 232—237, No. 8, April 3.

STARK. Die negative und die positive Elektrizitat in der Spectralanalyse. *Jahrb. d. Radioakt.*, **4**, 231, Sept. Vortrag, gehalten auf dem Nederlandsch Natuur-en Geneeskundig Congres in Leiden, April 6.

WOOD. Anomalous Dispersion and Ionisation. *Nature*, **75**, 583, April 18.

CARTER. The Fluorescence, Absorption, and Magnetic Rotation Spectra of Potassium Vapor. Abstract of a paper presented at the Washington Meeting of the Phys. Soc., April 19-20; *The Phys. Rev.*, **24**, 536.

PURVIS. The Influence of a Strong Magnetic Field on the Spark Spectra of Lead, Tin, Antimony, Bismuth, and Gold. *Cambridge Phil. Soc.*, May 6; *Nature*, **76**, 166.

WIEN. Über die absolute, von positiven Ionen ausgestrahlte Energie und die Entropie der Spektrallinien. *Ann. d. Phys.* (4), **23**, 415, May 9.

J. BECQUEREL. Sur les changements des bandes d'absorption des cristaux et la loi de variation de l'amortissement du mouvement des électrons absorbants à diverses températures. *Compt. rend.*, **144**, May 13.

J. BECQUEREL. Influence des variations de température sur les phénomènes d'absorption dans les cristaux. Phénomènes magnéto-optiques à la température de l'air liquide. *Soc. Franç. de Phys.*, June 7.

GEIGER. Über die Begleiterscheinung des inversen longitudinalen Zeeman effektes. *Ann. d. Phys.* (4), **23**, 758, July 6; *Inaug. Diss.*, Göttingen.

J. BECQUEREL. Untersuchungen über die magneto-optischen Erscheinungen in Kristallen. *Phys. Z.S.*, **8**, 632, No. 19, July 16.

COTTON AND MOUTON. Nouvelle propriété optique (biréfringence magnétique) de certains liquides organiques non colloïdaux *Compt. rend.*, **145**, 229, July 22; *Soc. Franç. de Phys.*, No. 268, 3.

PURVIS. Untersuchungen über die Zeeman phänomene. *Phys. Z.S.*, **8**, 594, July.

ANNA STETTENHEIMER. Eine absolute Messung des Zeeman phänomens. *Ann. d. Phys.*, (4), **24**, 384; *Inaug. Diss.*, July.

PASCHEN. Über die absolute Messung des Zeeman-Effektes. *Phys. Z.S.*, **8**, 522, July.

WOOD. The Magnetic Rotation of Sodium Vapour at the D-Lines. *Phil. Mag.*, (6), **14**, 145, July.

MILLER. Zeeman effekt an Magnesium, Calcium, Strontium, Zink, Cadmium, Mangan, und Chrom. *Ann. d. Phys.*, (4), **24**, 105, Aug. 9; *Inaug. Diss.*, Göttingen.

MÜLLER. Das optische Verhalten der kolloïdalen Metalle. *Ann. d. Phys.*, (4), **24**, 1, Aug. 10.

J. BECQUEREL. Sur les variations des bandes d'absorption des cristaux de parisite et de tysonite dans un champ magnétique, à la température de l'air liquide. *Compt. rend.*, **145**, 413, Aug. 19.

J. BECQUEREL. Influence des variations de température sur l'absorption dans les corps solides. *Le Radium*, Sept.

VOIGT. Betrachtungen über die Komplizierten Formen des Zeeman-effektes. *Ann. d. Phys.*, (4), **24**, 193, Sept.

GEIGER. Über die Begleiterscheinungen des inversen longitudinalen Zeeman-effektes. *Ann. d. Phys.*, (4), **24**, 597, Oct. 10.

ZEEMAN. The Intensities of the Components of Spectrum Lines divided by Magnetism. *Proc.* Roy. Acad. Amsterdam, **10**, 289, Oct. 26; *Versl.* K. Ak. van Wet. (Afd. Nat.), **16**, 286; *Le Radium*, **5**, No. 2, 49, Feb., 1908.

J. BECQUEREL. Sur la mesure de la dispersion anomale dans les cristaux à diverses températures et sur quelques conséquences théoriques. *Compt. rend.*, **145**, 795, Nov. 11.

J. BECQUEREL. Dispersion rotatoire magnétique des cristaux aux environs des bandes d'absorption. *Compt. rend.*, **145**, 916, Nov. 25.

HARTMANN. Das Zeeman-Phänomen im sichtbaren Spektrum von Kupfer, Eisen, Gold und Chrom. *Inaug. Diss.*, Halle-Wittenberg, Nov.

J. BECQUEREL. L'emploi de très basses températures pour l'analyse spectrale et pour l'étude des phénomènes magnéto-optiques dans les solutions. *Compt. rend.*, **145**, 1150, Dec. 9.

ZEEMAN. Waarneming van de magnetische splitsing der spectraallijnen met de methode van Fabry en Perot. (Observations of the Magnetic Resolution of Spectrum Lines by Means of the Method of Fabry and Perot.) *Versl.* K. Ak. van Wet. (Afd. Nat.), Dec. 28, 486; *Proc.* Roy. Acad. Amsterdam, 440, Dec. 28; *Phys. Z.S.*, 9 Jahrg., No. 7, S. 209–212.

J. BECQUEREL. Variations des bandes d'absorption des sels de didyme et d'erbium dans un champ magnétique. *Compt. rend.*, **145**, 1412, Dec. 30.

WIEN. Gesetze und Theorien der Strahlung. *Sitzber. der Physikal-med. Gesellschaft zu Würzburg.*

ERFLE. Optische Eigenschaften und Elektronen theorie. *Diss. München*, Techn. Hochschule.

DREPPER. Über anomale Dispersion der magnetischen Rotationspolarisation. *Diss. München.*

GEIGER Beiträge zur Kenntnis der Begleiterscheinung des inversen longitudinalen Zeeman-effektes. *Diss.* Leipzig (58 S.).

WEINHOLD. Elektronen theorie des Lichts. Teilung der Spektrallinien im Magnetfeld. *Bericht der Naturwissenschaftlichen Gesellschaft zu Chemnitz*, **16**, (1903–1907), 1907, **37–39**.

1908.

MOORE. Über die Zerlegung der Spektrallinien von Barium, Yttrium, Zirkon und Osmium im magnetischen Felde. *Ann. d Phys.*, **25**, S. 309, Jan. 19.

J. BECQUEREL. Recherches sur les phénomènes magnéto-optiques dans les cristaux et les solutions solidifiées à la température de l'air liquide. *Le Rad um*, Jan.

GMELIN. Uber die unsymmetrische Zerlegung der gelben Quecksilberlinie 5790 im magnetischen Felde. *Phys. Z.S.*, 9 Jahrg., No. 7, 212, Feb. 24.

DU BOIS AND ELIAS. Der Einfluss von Temperatur und Magnetisierung bei selektiven Absorptions- und Fluoreszenzspektren. *Versl.* K. Ak. van Wet. (Afd. Nat.), Feb. 29, March 28, April 24; *Verh. d. D. Physik. Ges.* 10 Jahrg., No. 12, May 15; *Ann. d. Phys.*, **27**, S. 233, July 31; *Arch. Néerl.*, (2), **14**, 87.

ZEEMAN. Nieuwe waarnemingen over asymmetrisch door magnetisme gesplitste tripletten. (New Observations Concerning Asymmetrical Triplets.) *Versl.* K. Ak. van Wet., 610, Feb. 29; *Proc.* Roy. Acad. Amsterdam, 566, Feb. 29; *Phys. Z.S.*, 9 Jahrg., No. 10, 340.

ZEEMAN. Verandering van golflengte van de middelste lijn van tripletten in een magnetisch veld I, II. (Change of Wave-length of the Middle Line of Triplets, I, II.) *Versl.* K. Ak. van Wet. (Afd. Nat.), 618, Feb. 29; 855, April 24; *Proc.* Roy. Acad. Amsterdam, 574, Feb. 29; 862, April 24; *Phys. Z.S.*, 10 Jahrg., No. 7, 217; *Arch, Néerl.*, Séries 2, **14**, 267.

J. BECQUEREL AND KAMERLINGH ONNES. Sur l'absorption de la lumière et les phénomènes magnéto-optiques dans les sels de terres rares, aux temperatures de liquéfaction et de solidification de l'hydrogène. *Versl.* K. Ak. van Wet. (Afd. Nat.), Feb. 29; *Soc. Franç. de Phys.*, June 5; *Compt. rend.*, March 23.

WOOD. Anomalous Magnetic Rotatory Dispersion of Neodymium. *Phil. Mag.*, Series 6, **15**, Feb.

WOOD. On the Existence of Positive Electrons in the Sodium Atom. *Phil. Mag.*, Series 6, **15**, Feb.

NATANSON. On the Elliptic Polarisation of Light Transmitted through an Absorbing Gaseous Medium, Parallel to the Lines of an Extraneous Magnetic Field. *Bull.* de l'ac. de Crac., Séance du March 2.

DUFOUR. Sur quelques exemples de raies présentant le phénomène de Zeeman anormal dans le sens des lignes de force magnétique. *Compt. rend.*, March 23.

DU BOIS, ELIAS, AND LÖWE. Een autocollimatieve spektraaltoestel van groote lichtintensiteit, tevens monochromator. (An Auto-collimating Spectrum Apparatus of great Luminous Intensity.) *Versl.* K. Ak. van Wet (Afd. Nat.), March 28; *Proc.* Roy. Acad. Amsterdam, March 28, 729.

J. BECQUEREL. Sur un phénomène attribuable à des électrons positifs, dans le spectre d'étincelle de l'yttrium. *Compt. rend.*, March 30.

ELIAS. Über anomale magnetische Rotationspolarisation der seltenen Erden. *Phys Z.S.*, 9 Jahrg., No 11, S. 355, April 16.

v. BAYER. Über den Zeemaneffekt in schwachen Magnetfeldern. *Verh. d. D. physik. Ges.*, 10 Jahrg., No. 10, May 1.

DUFOUR. Action du champ magnétique sur les bandes des spectres d'émission de quelques corps composés à l'état gazeux. *Soc. Franç. de Phys.*, June 5.

LAUE. Das Zeeman-Phänomen und der zweite Hauptsatz der Thermodynamik. *Phys. Z S.*, 9 Jahrg., No. 19, S. 617–620, June 11 and S. 907.

v. BAYER AND GEHRCKE. Uber den Zeeman effekt in schwachen Magnetfeldern. *Verh. d. D. physik. Ges.*, 10 Jahrg., No. 12, June 19.

HALE. Solar Vortices and the Zeeman Effect. *Publications* of the Astronomical Society of the Pacific, **20**, 200, July 3.

COTTON AND MOUTON. Sur l'orientation des cristaux pas le champ magnétique. Importance, au point de vue de la symétrie cristalline, des propriétés optiques des liqueurs mixtes. *Compt. rend.*, **147**, 51, July 6.

COTTON AND MOUTON. Biréfringence magnétique et électrique de la nitrobenzine. Variation avec la longueur d'onde. *Compt. rend.*, **147**, 193, July 20.

BEHRENS. Über magnetische Zirkularpolarisation in Metallen. *Inaug. Diss.*, Münster, July.

JACK. Zeeman-Effekt an Wolfram und Molybdän. *Inaug. Diss.*, Göttingen, July. *Ann. d. Phys.*, (4), **28**, 1032.

J. BECQUEREL. Sur les électrons positifs. *Le Radium*, July.

J. BECQUEREL. On the Dispersion of Magnetic Rotatory Power in the Neighbourhood of Bands of Absorption in the Case of Rare Earths. Remarks on the Subject of Recent Communications by Prof. R. W. Wood and Mr. G. J. Elias. *Phil. Mag.*, July.

ZEEMAN. Solar Magnetic Fields and Spectrum Analysis. *Nature*, 369, Aug. 20.

ZEEMAN. Spektralanalytische Untersuchung der magnetischen Felder auf der Sonne. *Ver. d. D. physik. Ges.*, 10 Jahrg., Nos. 18–20, Sept. 23; *Phys. Z.S.*, 9 Jahrg., No. 23, S. 834.

VOIGT. Eine Einwirkung natürlich aktiver Körper auf das an ihnen reflektierte Licht (nach Beobachtungen von K. Forsterling). *Verh. d. D. physik. Ges.*, 10 Jahrg., Nos. 18–20, Sept. 23.

MRS. VAN BILDERBEEK-VAN MEURS. The Zeeman Effect of the Strong Lines of the Violet Spark Spectrum of Iron in the Region of λ 2380—λ 4416. *Proc.* Roy. Acad. Amsterdam, Sept. 26.

HALE. Solar Vortices. Contributions from the Mount Wilson Solar Observatory, No. 26; *Astrophys. Journ.*, **28**, 100, Sept.

HALE. On the Probable Existence of a Magnetic Field in Sun-spots. Contributions from the Mount Wilson Solar Observatory, No. 30, Oct. 7; *Astrophys. Journ.*, **28**, Nov.; Terrestrial Magnetism and Atmospheric Electricity, Dec.

JACK. Dissymmetrical Separations in the Zeeman Effect in Tungsten and Molybdenum. *Proc.* R. Edinb. Soc., **29**, Part 2, No. 5, 75, Oct. 12.

DUFOUR. Modifications normales et anormales, sous l'influence d'un champ magnetique, de certaines bandes des spectres d'émission de molécules de divers corps à l'état gazeux. *Le Radium*, **5**, 291, Oct.; *Bull. Soc. fr. d. Phys.*, 225; *Journ. de Phys.*, (4), **8**, 237.

NATANSON. On the Elliptic Polarization of Light Transmitted through an Absorbing Naturally-active Medium. *Bull. de l'ac. de Crac.*, Oct.

NAGAOKA AND AMANO. Zeeman Effect in Weak Magnetic Fields. Tôkyô Sûgaku-Buturigakkwai Kizi, Series 2, **4**, No. 21, 421, Nov. 21; *Phys. Z.S.*, **10**, 159; *Arch. sc. phys. et nat.* (4), **27**, 275.

PURVIS. The Radiation of Various Spectrum Lines of Neon, Helium, and Sodium in a Magnetic Field. *Proc. Cambr. Soc.*, **15**, Part 1, 45, Nov. 23; *Nature*, **79**, 178.

COTTON AND WEISS. Sur le rapport de la charge à la masse des électrons: Comparaison des valeurs déduites de l'étude du phénomène de Zeeman et de mesures récentes sur les rayons cathodiques. *Compt. rend.*, Nov. 23.

VOIGT. Remarks on the Leyden Observations of the Zeeman-effect at Low Temperatures. *Proc.* Roy. Acad. Amsterdam, Nov. 28.

J. BECQUEREL. Sur les modifications des spectres d'émission et d'absorption sous l'influence d'un champ magnétique et sur la question des électrons positifs. *Le Radium*, **5**, 356, Dec. 7.

MARIO TENANI. Il fenomeno di Zeeman et il secondo principio della termodinamica. *Lincei Rend.*, (5), **17**, [2], 714, Dec. 20.

P. ZEEMAN. De wet van verplaatsing van de middelste lijn van een triplet. (The Law of Shift of the Central Component of a Triplet in a Magnetic Field.) *Versl.* K. Ak. Amsterdam, 541, Dec. 24; *Proc.* Roy. Acad. Amsterdam, 473; *Phys. Z.S.*, **10**, No. 7, 220; *Arch. Néerl.*, Séries 2, **14**, 273.

H. NAGAOKA. Zeeman Effect in Weak Magnetic Fields. *Nature*, **79**, 221, Dec. 24.

A. DUFOUR. Sur le pouvoir rotatoire magnétique de la vapeur de fluorure de calcium et de la vapeur d'hypoazotide au voisinage de leurs bandes d'absorption. *Compt. rend.*, **147**, 1471, Dec. 28.

P. ZEEMAN. Recherches sur la décomposition magnétique des raies spectrales. *Arch. Néerl.*, Séries 2, **13**, 260.

1909.

O. v. BAEYER. Bericht über die Struktur der Spektrallinien. *Jahrb. d. Rad.*, **6**, S. 50 and 245.

J. BECQUEREL. Sur quelques phénomènes optiques et magnéto-optiques dans les cristaux aux basses températures. *Compt. rend.*, **148**, 158, Jan. 18.

H. NAGAOKA. On the Complex Structure of some Spectrum Lines of Mercury. *Publ.* Tokyo Mathematico-phys. Soc. (2), **5**, No. 1, 2, Jan. 23.

L. NATANSON. Note Supplementary to a Paper On the Elliptic Polarisation of Light Transmitted through an Absorbing Naturally-active Medium. *Bull. de l'ac. de Crac.*, 25, Jan.

A. S. KING. On the Separation in the Magnetic Field of some Lines Occurring as Doublets and Triplets in Sun-spot Spectra. *Astrophys. Journ.*, **29**, Jan.; Contributions from the Wilson Solar Observatory, No. 34.

O. v. BAEYER. Eine Methode zur Vermehrung der Helligkeit und Schärfe der Interferenzstreifen an planparallelen Platten. *Verh. D. Phys. Ges.*, **11**, 118, Feb. 5.

A. DUFOUR. Sur les phénomènes de Zeeman normaux et anormaux. *Le Radium*, **6**, 44, Feb. 13.

P. GMELIN. Der Zeemaneffekt einiger Quecksilberlinien in schwachen Magnetfeldern. Absolut gemessen. *Ann. der Phys.*, (4), **28**, 1079, Feb. 15; *Inaug. Diss.*, Tubingen.

P. ZEEMAN. Änderung der Wellenlänge an der Mittellinie von Tripletts. *Phys. Z.S.*, **10**, No. 7, 217, Feb. 22; *Arch. Néerl.*, (2), **14**, 267.

B. E. MOORE. Der Zeemaneffekt bei den Linien des Thoriumspektrums. *Phys. Z.S.*, **10**, 297, March 2.

W. RITZ. Über einige anomale Zeemaneffekte im Spektrum von Thorium (Bemerkung zu der Arbeit des Herrn Moore). *Phys. Z.S.*, **10**, 307, March 2.

H. A. WILSON. An Attempt to Detect some Electro-optical Effects. *Proc.* Roy. Soc., (A), **82**, 224. March 5.

J. BECQUEREL. Sur la question de l'émission et de l'absorption de la lumière incomplètement polarisée dans un champs magnétique et sur le phénomène de Zeeman dans les spectres cannelés. *Compt. rend.*, **148**, 707, March 15.

G. J. ELIAS. Über anomale Dispersion der magnetischen Rotationspolarisation (Zweite Mitteilung). *Verh. D. Phys. Ges.*, **11**, 207, March 19.

A. DUFOUR. Sur les phénomènes de Zeeman normaux et anormaux dans les spectres des vapeurs. *Compt. rend.*, **148**, 775, March 22.

F. KOLÁČEK. Phänomenologisches über Dispersion und Zeeman effekt. *Ann. d. Phys.*, (4), **29**, 466, March 29.

R. W. WOOD. The Complete Balmer Series in the Sodium Spectrum. *Astrophys. Journ.*, **29**, 97, March.

A. DUFOUR. Sur l'hypothèse de l'existence d'électrons positifs dans les tubes à vide. *Bull. Soc fr. de Phys.*, 4th fasc., 36, April 2.

A. DUFOUR. Sur quelques expériences récentes relatives à l'hypothèse de l'existence d'electrons positifs aux basses pressions dans certains tubes à décharges électriques. *Le Radium*, **6**, 117, April 4. *Bull. Soc. fr. de Phys.*, 1st fasc., 61.

J. BECQUEREL. Sur un type nouveau de décomposition magnétique des bandes d'absorption des cristaux. Production simultanée de systèmes polarisés circulairement en des sens opposés. *Compt. rend.*, **148**, 913, April 5.

G. J. ELIAS. Het Zeeman effect bij selectieve absorptie en fluorescentie spektra. (The Zeeman Effect of Selective Absorption and Fluorescence Spectra.) Hand. XII, Ned. Nat. et Gen. Congr. Utrecht, 125, April 17.

G. E. HALE. Sur les champs magnétiques des taches solaires. *Bull. Soc. fr. de Phys.*, 118, April 23; *Journ. de Phys.*, (4), **8**, 461.

C. Barus. On an Adjustment for the Plane Grating similar to Rowland's Method for the Concave Grating. *Proc.* Amer. Phil. Soc., **48**, 166, April 24.

A. Cotton. Remarques sur le phénomène de Zeeman. *Bull. Soc. fr. de Phys.*, 4th fasc., 55, May 7.

G. E. Hale. Some Recent Contributions to our Knowledge of the Sun. Hamilton Lecture. From *Smithsonian Misc. Collections*, **52**, Part 3, 331 May 8.

A. Dufour. Sur un exemple de phénomène de Zeeman longitudinal positif pur dans les spectres d'émission des vapeurs. *Compt. rend*, **148**, 1311, May 17.

G. J. Elias. Anomale magnetische draaiingsdispersie in verband met selectieve absorptie (Anomalous Magnetic Rotatory Dispersion in Connexion with Selective Absorption. *Diss. Utrecht*, May 28.

H. Nagaoka. On the Zeeman Effect. *The Astronomical Herald*, published by the Astronomical Society of Japan, **2**, No. 2, May (Japanese).

Mario Tenani. Sulla scomposizione magnetica delle linee spettrali. *Lincei Rend.*, (5), **18**, [1], 595, June 5.

O. v. Baeyer and E. Gehrcke. Über den Zeemaneffekt in schwachen Magnetfeldern. *Ann. d. Phys.*, (4), **29**, 941, June 8.

H. Stansfield. The Echelon Spectroscope, its Secondary Action and the Structure of the Green Mercury Line. *Phys. Soc.*, London, June 11; *Nature*, **80**, 509; *Phil. Mag.*, (6), **18**, 371. *Proc.* Phys. Soc., London, **21**, 822.

A. Dufour. Observation, faite parallèlement aux lignes de force, des dissymétries de positions et d'intensités des composantes magnétiques de certaines raies d'émission; nouveau type de dissymétrie de positions. *Compt. rend.*, **148**, 1594, June 14.

R. W. Wood. Absorption, Fluorescenz, magnetische Rotation und anomale Dispersion des Quecksilberdampfes. *Phys. Z.S.*, **10**, 466, June 16; *Phil. Mag.* (6), **18**, 240.

Mario Tenani. Sul comportamento magnetico-ottico della linea b 4 del Magnesio. *Lincei Rend.*, (5), **18**, [1], 677, June 20.

J. Becquerel. Contribution à la connaissance du phénomène de Zeeman dans les cristaux. *Versl.* K. Akad. Amsterdam, **18**, 146, June 26; Onnes Comm. Leiden Suppl., No. 20 to 109–120.

H. Du Bois. Een verbeterde halfring electromagneet. (An Improved Semicircular Electromagnet.) *Versl.* K. Akad. Amsterdam, **18**, 118, June 26; *Proc.* Amsterdam, **12**, 189.

H. A. Lorentz. Over de theorie van het Zeeman-effect bij waarneming in willekeurige richting. (On the Theory of the Zeeman Effect in a Direction Inclined to the Lines of Force.) *Versl.* K. Akad. Amsterdam, **18**, 126, June 26; *Proc.* Amsterdam, **12**, 321; *Arch. Néerl.*, (2), **15**, 429.

Loria. Bestimmung der Dispersion des Lichtes in nicht leuchtendem, gesättigtem Natriumdampfe von der Temperatur 385° C. *Bull. de l'ac. de Crac*, 39, June.

A. H. Pfund. A Simple Interferometer for Showing the Zeeman Effect. *Phys. Rev.*, **28**, 446, June.

J. Becquerel. Sur l'existence dans la décomposition magnétique des bandes d'absorption d'un cristal uniaxe de dissymétries de position observées parallélement aux lignes de force du champ et à l'axe optique du cristal. *Compt. rend.*, **149**, 200, July 19.

A. S. King. The Zeeman Effect for Titanium. *Astrophys. Journ.*, **30**, 1, July; Contributions from the Mt. Wilson Solar Observatory, **39**, 97.

A. Cotton and H. Mouton. Variation, avec la température, de la biréfringence magnétique des composés aromatiques. Corps surfondus et corps à l'état vitreux. *Compt. rend.*, **149**, 340, Aug. 2.

H. Nagaoka. Über die Struktur der Spektrallinien des Quecksilbers. *Phys. Z.S.*, **10**, 609, Aug. 2.

J. Becquerel. Sur différentes espèces de dissymétries d'intensités, observées pour les composantes magnétiques, polarisées circulairement, des bandes d'absorption des cristaux uniaxes. *Compt. rend.*, **149**, 392, Aug. 9.

H. Nagaoka. Difference between Longitudinal and Transversal Zeeman Effects in Helium Lines. *Nature*, **81**, 188, Aug. 12 ; *Proc.* Tokyo Math. Phys. Soc., (2), **5**, 144.

J. Stock. Über asymmetrische Tripletts im Wolframspektrum. *Phys. Z.S.*, **10**, 694, Aug. 26.

F. Paschen. Über die Seriensysteme in den Spektren von Zink, Cadmium und Quecksilber. *Ann. d. Phys.*, (4), **30**, 746, Sept. 16.

B. E. Moore. Upon the Separation of the Spectrum Lines of Thorium in the Magnetic Field. *Astrophys. Journ.*, **30**, 144, 178, Sept.

G. E. Hale. Note on the Magnetic Field in Sun-spots. *Publications* Astron. Soc. of the Pacific, **21**, No. 128, 205, Oct. 10.

P. Zeeman. De graad van volledigheid der circulaire polarisatie van magnetisch gesplitste lijnen. (The Degree of Completeness of the Circular Polarisation of Magnetically Divided Lines.) *Versl.* K. Akad. Amsterdam, **18**, 317, Oct. 30 ; *Proc.* Amsterdam, **12**, 345 ; *Phys. Z.S.*, **11** (1910), S. 1 ; *Arch. Néerl.*, (2), **15** 179, (1910).

R. W. Wood. The Ultra-violet Absorption, Fluorescence, and Magnetic Rotation of Sodium Vapour. *Phil. Mag.*, (6), **18**, 530, Oct. ; *Phys. Z.S.*, **10**, 913.

J. Becquerel. Dissymétries de positions et d'intensités présentés par les composantes magnétiques polarisées circulairement, des bandes d'absorption d'un cristal uniaxe. Dissymétries d'intensités de deux natures différentes. *Le Radium*, **6**, 327, Nov. 6.

A. Dufour. Dissymétries dans le phénomène de Zeeman présenté par certaines bandes d'emission des vapeurs. *Compt. rend.*, **149**, 917, Nov. 22.

T. Royds. Anhang zur Arbeit von Hrn. F Paschen uber "Die Seriensysteme in den Spektren von Zink, Cadmium und Quecksilber." *Ann. d. Phys.*, (4), **30**, 1024, Nov. 23.

A. Cotton. Remarques théoriques sur le phénomène de Zeeman. *Bull. Soc. Franç. de Phys.*, 4th fasc , 83, Dec. 3.

A. Dufour. Sur les phénomènes de Zeeman dissymétriques rencontrés dans les spectres d'émission. *Bull. Soc. Franç. de Phys.*, 4th fasc., 82, Dec. 3.

A. Dufour. Dissymétrie de positions des composantes magnétiques rencontrée dans les spectres de bandes d'émission. *Le Radium*, **6**, 366, Dec. 7.

H. B. van Bilderbeek-van Meurs. Magnetische splitsing van het ultraviolette ijzerspectrum (λ 2300-λ 4500). (La décomposition magnétique des raies du spectre ultra-violet du fer). *Diss. Amsterdam*, Dec. 9 ; *Arch. Néerl.*, (2), **15**, 353 (1911).

J. Becquerel. Remarques au sujet des communications de M. Dufour et M. Cotton. *Bull. Soc. Franç. de Phys.*, 4th fasc., 87, Dec. 17.

J. Laub. Zur Theorie der longitudinalen magneto-optischen. Effekte in leuchtenden Gasen und Dämpfen. *Sitzungsber. Heidelberger Akad. d. Wiss.*, 6th Abh., Dec. 22.

J. Becquerel. Influence d'un champ magnétique sur l'amortissement des vibrations lumineuses. *Compt. rend.*, **149**, 1364, Dec. 27.

R. W. Wood and F. E. Hackett. The Resonance and Magnetic Rotation Spectra of Sodium Vapour Photographed with the Concave Grating. *Astrophys. Journ.*, **30**, 339, Dec.

O. M. Corbino. L'emissione luminosa nei vari azemut da parte d'un vapore incandescente in un campo magnetico. *Cim.* (5), **17**, 144.

H. A. Lorentz. Theorie der magneto-optischen Phanomene. *Encykl. der Math. Wiss.*, **5**, 3, Heft 2, 199.

A. A. Michelson. Recent Advances in Spectroscopy. Nobel Lecture. Les prix Nobel en 1907.

Curt Peterke. Untersuchungen über das Zeeman-Phänomen in sichtbaren Spektrum von Nickel und Palladium. *Diss.*, Halle a. S.

O. D. Tauern. Über das Auftreten des Kerrphanomens in Gläsern. *Diss.*, Freiburg i. Br.

P. Zeeman. L'origine des couleurs du spectre. *Rivista di Scienza*, "Scientia," **5**.

1910

G. Meslin. Dichroisme magnétique des liqueurs constituées par la sidérose et orientation des cristaux de sidérose dans le champ. *Bull. Soc. Franç. de Phys.*, 54.

P. ZEEMAN AND B. WINAWER. De magnetische splitsing van absorptielijnen in verband met het spectrum der zonnevlekken. (The Magnetic Separation of Absorption Lines in Connexion with Sun-spot Spectra). *Versl.* K. Acad. Amsterdam, **18**, 621 (Jan. 29); 889 (April 29); **19**, 233, June 25; *Proc.* Amsterdam, **12**, 584, **13**, 35, 162; *Phys. Z.S.*, **11**, 553, 657, 880; *Arch. Néerl.*, (2), **15**, 453. *Atrophys. Journ.*, **32**, 329.

MARIO TENANI. Sul compartamento magnetico-ottico di alcune, linee spettrali. *Lincei Rend.*, (5), **19**, [1], 198, Feb. 20.

C. FABRY. Production de champs magnétiques intenses au moyen de Bobines sans fer. *Journ. de Phys.*, (4), **9**, 129.

A. COTTON AND H. MOUTON. Sur la biréfringence magnétique des liquides purs. Comparaison avec le phénomène electro-optique de Kerr. *Ann. chim. phys.*, (8), **19**, 153, Feb.; **20**, 194; *Bull Soc. Franç. de Phys.*, 189; *Journ. de Phys.*, (5), **1**, 5.

MARIO TENANI. Sulla origine di alcune gravi anomalie recentemente osservate nello studio del fenomeno di Zeeman e su un nuovo metodo per lo studio di un campo magnetico. *Lincei Rend.*, (5), **19**, [1], 554, May 1.

O. M. CORBINO. Sulla distribuzione delle linee isodinamiche tra i poli di un elettromagnete. *Lincei Rend.*, (5), **19**, [1], 540, May 1.

A. DUFOUR. Sur les triplets magnétiques dissymétriques. *Le Radium*, **7**, 74, 153, March 7.

H. NAGAOKA. Longitudinal and Transversal Zeeman Effects in D-Lines of Sodium and Green Line of Mercury. *Proc.* Tokyo Math.-Phys. Soc., (2), **5**, 272, March 19.

A. DUFOUR. Sur les triplets dissymétriques. Exemple d'une dissymétrie de position proportionnelle au carré du champ magnétique. *Compt. rend.*, **150**, 614, March 7.

A. COTTON AND H. MOUTON. Sur la biréfringence magnétique et électrique des liquides aromatiques et sur la theorie de l'orientation moléculaire. *Compt. rend.*, **150**, 774, March 21.

O. M. CORBINO. Sulle anomalie del fenomeno Zeeman. *Lincei Rend.*, (5), **19**, [1], 687, May 22.

O. M. CORBINO. Sul fenomeno Majorana nel campi oscillatori. *Cim.*, (5), **19**, [1], 375, May.

A. EAGLE. On a New Mounting for a Concave Grating. *Astrophys. Journ*, **31**, 120, March.

L. TIERI. Il fenomeno Majorana in campi magnetici prodotti da scariche oscillatorie. *Lincei Rend.*, (5), **19**, [1], 377, April 3.

O. M. CORBINO. Sulla origine di alcune gravi anomalie recentemente osservate nello studio del fenomeno Zeeman. *Lincei Rend.*, (5), **19**, [1], 466, April 17; *Cim.*, (5), **19**, [1], 425.

A. DUFOUR. Dissymétries dans le phénomène de Zeeman présenté par certaines raies et par certaines bandes des spectres d'émission des vapeurs. *Journ. de Phys.*, (4), **9**, 277, April. *Bull. Soc. Franç. de Phys.*, 80.

G. E. HALE. Notes on Solar Magnetic Fields and Related Phenomena. *Publ.* Astron. Soc. of the Pacific, **22**, No. 131, April.

H. NAGAOKA. Zeeman Effect of the Yellow Mercury Line $\lambda = 5770$. *Nature*, **83**, 276, May 5.

H. NAGAOKA. On a New Type of Nonet. *Proc.* Tokyo Math.-Phys. Soc., (2), **5**, 284, May 7.

H. NAGAOKA AND T. TAKAMINE. A Difference in the Change of Frequency between Longitudinal and Transversal Zeeman Effects. *Proc.* Tokyo Math.-Phys. Soc., (2), **5**, 278, May 7.

O. M. CORBINO. Die verteilung der Isodynamen zwischen den Polen eines Elektromagnets und Anomalien des Zeeman-phanomens. *Phys. Z.S.*, **11**, 521, May 15; *Arch. sci. phys. et nat.*, (4), **29**, 577.

G. VIETH. Über die magnetische Drehung der Polarisationsebene in kristallinisch flussigen Substanzen. *Phys. Z.S.*, **11**, 526, May 27; *Diss.* Halle.

A. COTTON AND H. MOUTON. Sur la biréfringence magnétique des liquides aromatiques. *Bull. Soc. Franç. de Phys.*, 54, June 3.

L. NATANSON. On the Theory of Double Refraction induced by an Electric or a Magnetic Field. *Krak. Anz.*, (A), 256, June 6.

P. P. Koch. Bemerkung zu dem behaupteten Unterschied zwischen longitudinalem und transversalem Zeemaneffekt. *Phys. Z.S.*, **11**, 612, June 25.

A. Cotton and H. Mouton. Étude et emplois des mica quart d'onde et des appareils à pénombres pour diverses radiations monochromatiques. Note au mémoire : Sur la biréfringence magnétique des liquides purs, etc. *Ann. chim. phys.*, (8), **20**, 275, June.

J. Guadet. Sur une méthode d'observation et d'étude du phénomène de Zeeman. *Ann. chim. phys.*, (8), **20**, 283, June.

A. S. King. The Correspondence between Zeeman Effect and Pressure Displacement for the Spectra of Iron, Chromium, and Titanium. *Astrophys. Journ.*, **31**, 433, June ; Contributions from the Mount Wilson Solar Observatory, No. 46.

A. Dufour. Sur la rotation de l'arc à mercure dans un champ magnétique. Observation du phénomène de Doppler. *Compt. rend.*, **151**, 60, July 4.

L. Tieri. Esaltazione della birifrangenza del ferro Bravais di recente preparazione. *Lincei Rend.*, (5), **19**, [2], 49, July 17.

H. Nagaoka. Longitudinaler und transversaler Zeemaneffekt an den Quecksilberlinien 5461 und 5769 A.E. *Phys. Z.S.*, **11**, 789, July 30.

L. Tieri. Sulla birifrangenza magnetica. *Cim.*, (5), **20**, July 21.

O. M. Corbino. Über das Majoranaphänomen in oszillatorischen Feldern und die von der Brownschen Bewegung herruhrende, der Orientierung entgegenwirkende Kraft. *Phys. Z.S.*, **11**, 756, Aug. 6.

L. Mandelstam. Über die Dämpfung der Eigenschwingungen in leuchtendem Natriumdampf. *Phys. Z.S.*, **11**, 752, Aug. 6.

G. von Ubisch. Schwingungszahl und Dämpfung im leuchtenden und nicht leuchtenden Natriumdampf. *Phys. Z.S.*, **11**, 753, Aug. 6.

P. Langevin. Sur les biréfringences électrique et magnétique. *Compt. rend.*, **151**, 475, Aug. 16 ; *Le Radium*, **7**, 233.

Mario Tenani. Sulla natura delle particelle ultramicroscopiche che intervengono nel fenomeno Majorana e su un nuovo metodo di studio del campo magnetico. Replica al prof. O. M. Corbino. *Lincei Rend.*, (5), 19, [2], 178, Aug. 21.

P. Langevin. Sur les biréfringences électrique et magnétique. *Le Radium*, **7**, 249, Aug. 25.

H. Nagaoka. Phénomène de Zeeman longitudinal et transversal des raies D du sodium. *Arch. Sc. Phys. et Nat.*, (4), **30**, 121, Aug.

H. E. J. G. Du Bois. Een verbeterde halfring electromagneet, II. (An Improved Semicircular Electromagnet.) *Versl.* K. Acad. Amsterdam, **19**, 397, Sept. 24 ; *Proc.* Amsterdam, **13**, 386.

G. J. Elias. Over het Zeeman-effect bij emissielijnen in eene richting schuin ten opzichte van de krachtlijnen. (On the Zeeman Effect for Emission Lines in a Direction Oblique with regard to the Lines of Force.) *Versl.* K. Acad. Amsterdam, **19**, 402, Sept. 24 ; *Proc.* Amsterdam, **13**, 391.

J. Cabannes. Sur la biréfringence électrique des liquides homogènes. Études des retards absolus. *Le Radium*, **7**, 364, Nov. 5.

A. Cotton and H. Mouton. Mesure absolue de la biréfringence magnétique de la nitrobenzine. *Compt. rend.*, **151**, 862, Nov. 14.

P. Gmelin. Über das Verhalten der Quecksilberlinie 5789, 88 im magnetischen Felde. *Phys. Z.S.*, **11**, 1193, Nov. 26.

P. P. Koch. Über die Messung der Intensitätsverteilung in Spektrallinien, I. Mit Anwendungen auf Interferenzspektroscopie. *Ann. d. Phys.*, (4), **34**, 377, Dec. 3.

A. Garbasso. La trajettoria caratteristica del fenomeno de Zeeman. *Lincei Rend.*, (5), **19**, [2], 562, Dec. 4.

J. Becquerel. Sur l'effet magnéto-optique de sens positif présenté par les bandes de phosphorescence du rubis et de l'émeraude et sur les relations entre l'émission et l'absorption dans un champ magnétique. *Compt. rend.*, **151**, 1344, Dec. 27.

A. Dufour. Nouvelles mesures du phénomène de Zeeman présenté par quelques bandes d'émission de molécules de corps à l'état gazeux. *Ann. Chim. Phys.*, (8), **21**, 568, Dec.

W. F. Schulz. The Effect of a Magnetic Field upon the Absorption Spectra of certain Rare Earths. *Astrophys. Journ.*, **30**, 383, Dec.

H. Du Bois. L'effet Zeeman généralisé dans les absorbants sélectifs. Résumé des principaux faits positifs. Congrès Intern. de Radiologie et d'Electricité.

G. J. Elias. Brechung und magnetische Doppelbrechung von Lösungen seltener Erden. *Verh. D. Phys Ges.*, **12**, 955.

B. Seegert. Der Zeeman-Effekt. Prometheus, **21**, 280.

M. Thouvenot. Contribution à l'étude de la Polarisation rotatoire magnétique. Nancy, 1910.

1911

H. Du Bois. L'effet Zeeman généralisé dans les absorbants sélectifs. *Arch. Néerl.*, (3), **1**, (1912), 209 ; *Phys. Z.S.*, **13**, (1912), 128 (Auszug).

A. Dufour. Rotation spontanée et rotation dans un champ magnétique de l'arc à mercure. Observation du phénomène de Doppler. *Journ. d. Phys.*, (5), **1**, 109, Jan. 6 ; *Bull. Soc. franç. de Phys.*, 4th fasc., 86.

H. Lunelund. Über die Struktur einiger Spektrallinien und ihren Zeeman-effekt in schwachen Magnetfeldern. *Ann. d. Phys.*, (4), **34**, 505, Jan. 8.

W. Geoffrey Duffield. The Effect of Pressure upon Arc-spectra, No. 3—Silver λ 4000 to λ 4600 ; No. 4—Gold. *Phil. Trans. A.*, **211**, 33, Jan. 9.

A. Cotton. Mesures sur le phénomène de Zeeman Changements magnétiques des raies d'émission des corps à l'état gazeux. *Le Radium*, **8**, 33, Jan. 13.

J. Becquerel. Sur les modifications magnétiques des bandes de phosphorescence et d'absorption du rubis et sur une question fondamentale de magnéto-optique. *Compt. rend.*, **152**, 183, Jan. 23.

P. Zeeman. Beschouwingen over lichtstraling onder den gelijktijdigen invloed van electrische en magnetische krachten en eenige naar aanleiding daarvan genomen proeven, I. (Considerations Concerning Light Radiation under the Simultaneous Influence of Electric and Magnetic Forces and some Experiments thereby suggested, I.) *Versl.* K. Acad. Amsterdam, 19, 957, Jan. 28 ; *Proc.* Amsterdam, **14**, 2 ; *Phys. Z.S.*, **13**, 177.

A. Dufour. Sur le phénomène de Zeeman présenté par les groupes de raies des spectres du type II. *Le Radium*, **8**, 97, Feb. 1.

E. Darmois. Recherches sur la polarisation rotatoire naturelle et la polarisation rotatoire magnétique. *Ann. chim. phys.*, (8), **22**, 247, (Feb.) ; 495 (April).

A. Dufour. Rotation spontanée et rotation dans un champ magnétique de l'arc à mercure. Observation du phénomène de Doppler. *Ann. chim. phys.*, (8), **22**, 282, Feb.

W. Voigt. Zur Frage der Dissymetrie der Zeemanschen Triplets. *Ann. d. Phys.*, (4), **35**, 101, March 19.

F. Paschen. Erweiterung der Seriengesetze der Linienspektra auf Grund genauer Wellenlängenmessungen im Ultrarot. *Jahrb. Radioaktivität und Elektronik*, **7**, 174, March 20.

G. J. Elias. Anomale magnetische Drehungsdispersion und selektive Absorption. *Ann d. Phys.*, (4), **35**, 299, March 22.

A. Cotton. Détermination des lignes neutres d'une lame cristalline. Applications à divers appareils d'optique physique. *Ann. chim. phys.*, (8), **22**, 428, March.

A. Dufour. Sur la comparaison des écarts des doublets magnétiques observés parallèlement ou perpendiculairement au champ magnétique. *Journ. de Phys.*, (5), **1**, 189, March.

H. Lunelund. Über das Verhalten der Trabanten—0·121 Å.E. der Quecksilberlinie 5790 Å.E. im magnetischen Felde. *Phys. Z.S.*, **12**, 511, April 11.

P. Zeeman. Le cas général de la décomposition magnétique des raies spectrales et son application en astrophysique. *Journ. de Phys.*, (5), **1**, 442, April 19 ; *Phys. Z.S.*, **13**, 86 (1912).

B. E. Moore. Über die Trennung der Spektrallinien des Calciums und des Strontiums im Magnetfelde. *Phys. Z.S.*, **12**, 433, April 20.

H. D. Babcock. The Zeeman Effect for Chromium. *Astrophys. Journ.*, **33**, 217, April ; Contributions from the Mount Wilson Solar Observatory, No. 52.

F. Schwers. Densité, polarisation rotatoire magnétique et indice de réfraction des mélanges binaires. Nouvelles contributions à l'étude des solutions. *Journ. chim. phys.*, **9**, 325, May 20.

P. P. KOCH. Zur Frage der Dissymmetrie der Zeemanschen Triplets. Bemerkung zu einer Veroffentlichung des Herrn Voigt. *Ann. d. Phys.*, (4), **35**, 1034, June 15.

H. DU BOIS AND G. J. ELIAS Der Einfluss von Temperatur und Magnetisierung bei selektiven Absorptions- und Fluoreszenzspektren (Zweite Mitteilung). *Ann. d. Phys.*, (4), **35**, 617.

· B. E. MOORE. On the Magnetic Separation of the Spectrum Lines of Calcium and Strontium. *Astrophys. Journ.*, **33**, 385, June.

H. DU BOIS AND G. J. ELIAS. Der Einfluss von Temperatur und Magnetisierung bei selektiven Absorptions- und Fluoreszenzspektren. *Verh. D. Phys. Ges.*, **13**, 345.

A. HEURUNG. Untersuchungen über die magneto-optischen Effekte bei Chlor und Jod. *Ann. d. Phys.*, (4), **36**, 153, July 21.

A. COTTON. Dichroïsme circulaire et dispersion rotatoire. *Compt. rend.*, **153**, 245, July 24.

A. COTTON. La théorie de Ritz du phénomène de Zeeman. *Le Radium*, **8**, 363, Aug. 12.

A. COTTON. Sur la pureté des raies spectrales lorsque l'intensité varie rapidement. Note au travail : "La théorie de Ritz du phénomène de Zeeman." *Le Radium*, **8**, 404, Aug 12.

A. COTTON. Une théorie du phénomène de Zeeman : La théorie de Ritz. *Revue générale des Sciences*, 22nd year, 597, Aug. 15.

S. RYBÁR. Über die Zerlegung der Spektrallinien von Lanthan und Kobalt im magnetischen Felde. *Phys. Z.S.*, **12**, 889, Sept. 4.

A. COTTON. Magnetische Doppelbrechung reiner Flüssigkeiten Neue Untersuchung gemeinsam mit H. Mouton. *Verh. D. Phys. Ges.*, **13**, 766, Sept. 26 ; *Phys. Z.S.*, **12**, 953.

R. LEISER. Über die elektrische Doppelbrechung der Gase. *Verh. D. Phys. Ges.*, **13**, 903, Sept. 26 ; *Phys. Z.S.*, **12**, 955.

H. D. BABCOCK. The Zeeman Effect for Vanadium. *Astrophys. Journ.*, **34**, 209, Sept. ; Contributions from the Mount Wilson Solar Observatory, No. 55.

W. VOIGT. Zwei Antworten. *Ann. d. Phys.*, (4), **36**, 866, Oct. 1.

W. VOIGT. Zur Theorie der komplizierten Zeemanseffekt. *Ann. d. Phys.*, (4), **36**, 873, Oct. 11.

A. S. KING. The Influence of a Magnetic Field upon the Spark Spectra of Iron and Titanium. Summary of Results. *Astrophys. Journ.*, **34**, 225, Oct. ; Contributions from the Mount Wilson Solar Observatory, No. 56.

M. MZ. RISCO. La asimetria de los Tripletes de Zeeman. Anales de la sociedad española de Fisica y Quimica. Año IX, Tome noveno, No. 86, 263, Oct. ; *Phys Z.S.*, **13**, 137 (1912).

A. COTTON. L'effet Zeeman positif dans les gaz et la théorie de Ritz. *Le Radium*, **8**, 449, Nov. 7.

R. FORTRAT. Remarques sur la théorie de Ritz du phénomène de Zeeman. *Le Radium*, **8**, 456, Nov. 15.

G. WENDT. Untersuchungen an Quecksilberlinien. Struktur Veränderung der Linien und des Spektrums bei Verdünnung des Metalldampfes. Zeemaneffekt in schwachen und starken Feldern. *Ann. d. Phys.*, (4), **37**, 535, Nov. 23.

P. ZEEMAN AND C. M. HOOGENBOOM. Elektrische dubbele breking in nevels. (Electric Double Refraction in some Artificial Clouds and Vapours.) *Versl.* K. Akad. Amsterdam, **20**, 570, Nov. 25; **20**, 921, Jan. 27, 1912; **21**, 188, June 29, 1912; *Proc.* Amsterdam, **14**, 558, 786; **15**, 178 ; *Phys. Z.S.*, **13**, 913 (1912)

W. VOIGT. Allgemeines über Emission und Absorption in Zusammenhang mit der Frage der Intensitätsmessungen beim Zeeman-Effekt. Nach Beobachtungen von C. Forsterling. Mit einem Zusatz von H. A. Lorentz. *Gött. Nachr. Math. Phys. Kl.*, S. 71, Nov. 26.

H. D. BABCOCK. Note on the Grouping of Triplet Separations Produced by a Magnetic Field. *Astrophys. Journ.*, **34**, 288, Nov. ; Contributions from the Mount Wilson Solar Observatory, No. 57.

J. E. PURVIS. Note on the Zeeman Effect for Chromium. *Astrophys. Journ.*, **34**, 312, Nov.

J. M. GRAFTDYK. Magnetische Splitsing van het Nikkel-en Kobaltspectrum en van het Yzerspectrum (λ 4400—λ 6500). (Décomposition magnétique des spectres du Nickel, du Cobalt et du Fer.) *Diss.*, Amsterdam, Dec. 15; *Arch. Néerl.*, (3), **2**, 192.

R. W. WOOD. Bemerkungen zu der A. Heurungschen Arbeit: Untersuchungen über die magneto-optischen Effekte bei Chlor und Jod. *Ann. d. Phys.*, (4), **37**, 594, Dec. 18.

A. COTTON. À propos d'un travail de O. M. Corbino, intitulé: Sur l'observation spectroscopique de la lumière dont l'intensité varie periodiquement. *Le Radium*, **9**, 19, Dec. 30 (1912).

O. LUTTIG. Das Zeeman-Phänomen im sichtbaren Spektrum von Mangan und Argon. *Diss.*, Halle a. S.

B. WINAWER. Zjarvisko Zeemana w świetle badań nowych. *Wiadomości Matematycznych*, **15**, 133.

1912

W. VOIGT AND H. M. HANSEN. Das neue Gitterspectroskop des Göttinger Institutes und seine Verwendung zur Beobachtung der magnetischen Doppelbrechung im Gebiete von Absorptionslinien. *Phys. Z.S.*, **13**, 217, Jan. 24.

L. CHAUMONT. Construction et vérification d'un quart d'onde à lame de mica. *Compt. rend.*, **154**, 271, Jan. 29.

H. J. V. BROTHERUS. Experimentelle und theoretische Untersuchungen über die Spektrallinien D_1 und D_2. *Diss.* Finnland, Feb. 14: *Ann. d. Phys.*, (4), **38**, 397.

O. LÜTTIG. Das Zeeman phänomen von Kupfer, Eisen, Gold, Chrom, Nickel, Palladium, Mangan und Argon im sichtbaren Spektrum. Nach Untersuchungen von W. Hartmann, C. Peterke und O. Lüttig zusammengestellt. *Ann. d. Phys*, (4), **38**, 43, Feb. 16.

W. VOIGT. Photographies de la biréfringence magnétique des flammes. *Soc. franç. de Phys.*, No. 21, 5, Feb. 16.

F. PASCHEN. Die Asymmetrie des Zeemanschen Tripletts. *Phys. Z.S.*, **13**, 252, Feb. 21.

J. MOLYNEUX. Einige Versuche zum Nachweis einer Verschiebung der Mittelkomponenten von normalen Zeemanschen Tripletts im He-Spektrum. *Phys. Z.S.*, **13**, 259, Feb. 23.

A. COTTON. Sur les mesures du phénomène de Zeeman. *Journ de Phys.*, (5), **2**, 97, Feb.; *Astrophys. Journ.*, **35**, 213.

A. HEURUNG. Erwiderung auf die Bemerkungen von R. W. Wood zu meinen Untersuchungen über die magneto-optischen Effekte bei Chlor und Jod. *Ann. d. Phys.*, (4), **37**, 1046, March 9.

A. COTTON AND H. MOUTON. Biréfringence magnétique et constitution chimique. *Compt. rend.*, **154**, 818, March 25.

A. COTTON AND H. MOUTON. Nouveaux corps présentants la biréfringence magnétique. Anisotropie moleculaire et atomique. *Compt. rend.*, **154**, 930, April 9.

F. POCKELS. Note sur les retards absolus dans les biréfringences electrique et magnétique. *Le Radium*, **9**, 148, April 15.

A. S. KING. The Effect of Pressure upon Electric Furnace Spectra (sec. paper). *Astrophys. Journ.*, **35**, 183, April; Contributions from the Mount Wilson Solar Observatory, No. 60.

W. VOIGT. Über elektrische und magnetische Doppelbrechung I, II, III. *Gött. Nachr.*, S. 577, 832, May 18, October 26.

K. FÖRSTERLING. Zur Theorie des Zeemaneffektes in beliebiger Richtung. *Gött. Nachr.*, S. 611, May 18.

T. VAN LOHUIZEN. Reeksen in de spectra van Tin en Antimoon. (Series in the Spectra of Tin and Antimony.) *Versl.* K. Akad. Amsterdam, **21**, 7, May 25, *Proc.* Amsterdam, **15**, 31.

T. VAN LOHUIZEN. Translatie-reeksen in lijnen spektra. (Translation Series in Line Spectra.) *Versl.* K. Akad. Amsterdam, **21**, 138, May 25; *Proc.* Amsterdam, **15**, 156.

F. CROZE. Contribution à l'étude du phénomène de Zeeman dans les spectres de l'hydrogène et de l'azote. *Compt. rend.*, **154**, 1410, May 28.

R. LADENBURG. Über die anomale dispersion und die magnetische Drehung der polarisationsebene des leuchtenden Wasserstoffs, sowie uber die Verbreiterung von spektrallinien. *Ann. d. Phys.*, (4), **38**, 249.

R. JACK. Magnetic Resolution of the Spectrum Lines of Niobium. *Proc.* Royal Irish Academy, **30**, Sect. A, No. 3, 42, June 10.

CH. WALI-MOHAMMED. Untersuchungen über magnetische Zerlegung feiner Spektrallinien im Vakuumlichtbogen. *Ann. d. Phys.*, (4), **39**, 225; *Diss.*, Göttingen.

F. PASCHEN AND E. BACK. Normale und anomale Zeemaneffekte. *Ann. d. Phys.*, (4), **39**, 897, July 19.

J. C. MCLENNAN. On the Constitution of the Mercury Green Line $\lambda = 5641$ Å.U., and on the Magnetic Resolution of its Satellites by an Echelon Grating. *Proc.* Roy. Soc. London, (A), **87**, 269.

M. MZ. RISCO. La asimetria de los Tripletes de Zeeman. *Diss.*, Madrid.

A. S. KING. The Influence of a Magnetic Field upon the Spark Spectra of Iron and Titanium. Papers of the Mount Wilson Solar Observatory, **2**, Part I.

P. ZEEMAN. On the Polarisation Impressed upon Light by Traversing the Slit of a Spectroscope and some Errors resulting therefrom. *Proc.* Roy. Acad., Amsterdam, October, 599.

K. KÖRNER. Zufällige dissymetrien Zeemanscher Tripletts. *Phys. Z.S.*, **13**, 797.

J. F. SIRKS. Metingen over de ultraviolette magnetische draaiing in gassen. (Measures of the Ultra-Violet Magnetic Rotation of Gases.) *Diss.*, Leiden, Nov. 25.

H. NAGAOKA and T. TAKAMINE. The Constitution of Mercury Lines examined by an echelon grating and a Lummer-Gehrcke Plate. *Proc.* Physical Society, London, Dec. 15.

G. E. HALE. Preliminary Note on an Attempt to Detect the General Magnetic Field of the Sun. *Terrestrial Magnetism*, **17**, 173.

INDEX

(The numbers refer to pages.)